Analytic Computational Complexity

ACADEMIC PRESS RAPID MANUSCRIPT REPRODUCTION

Proceedings of the Symposium on Analytic Computational Complexity held by the Computer Science Department, Carnegie-Mellon University, Pittsburgh, Pennsylvania, on April 7–8, 1975.

Analytic Computational Complexity

Edited by

J.F. Traub

Departments of Computer Science
and Mathematics
Carnegie-Mellon University
Pittsburgh, Pennsylvania

Academic Press

New York San Francisco London 1976

A Subsidiary of Harcourt Brace Jovanovich, Publishers

ACADEMIC PRESS, INC.
111 Fifth Avenue, New York, New York 10003

United Kingdom Edition published by
ACADEMIC PRESS, INC. (LONDON) LTD.
24/28 Oval Road, London NW1

Library of Congress Cataloging in Publication Data

Symposium on Analytic Computational Complexity, Carnegie-Mellon University, 1975.
Analytic computational complexity.

"Proceedings of the Symposium on Analytic Computational Complexity, held by the Computer Science Department, Carnegie-Mellon University, Pittsburgh, Pennsylvania, on April 7-8, 1975."
Bibliography: p.
Includes index.
1. Numerical analysis–Data processing–Congresses.
2. Computational complexity–Congresses. I. Traub, Joe Fred, (date) II. Carnegie-Mellon University. Computer Science Dept. III. Title.
QA297.S915 1975 519.4 75-13086
ISBN 0–12–697560–4

PRINTED IN THE UNITED STATES OF AMERICA

CONTENTS

CONTENTS

LIST OF INVITED AUTHORS

Richard P. Brent
Computer Centre, Australian National University, Box 4, Canberra, ACT 2600, Australia

B. Kacewicz
University of Warsaw, Institute of Mathematical Machines, P.K. i N. p.850, Warsaw 00-901, Poland

H. T. Kung
Computer Science Department, Carnegie-Mellon University, Pittsburgh, Pennsylvania 15213

Robert Meersman
Department of Mathematics, University of Antwerp, Universiteitsplein 1, 2610 Wilrijk, Belgium

John R. Rice
Computer Science Department, Mathematical Sciences Building, Purdue University, Lafayette, Indiana 47907

M.H. Schultz
Computer Science Department, Dunham Laboratory, Yale University, New Haven, Connecticut 06520

J. F. Traub
Computer Science Department, Carnegie-Mellon University, Pittsburgh, Pennsylvania 15213

S. Winograd
IBM T. J. Watson Research Center, P.O. Box 218, Yorktown Heights, New York 10598

H. Woźniakowski
University of Warsaw, Institute of Mathematical Machines, P.K. i N. p.850, Warsaw 00-901, Poland

David Y. Y. Yun
IBM T. J. Watson Research Center, P.O. Box 218, Yorktown Heights, New York 10598

PREFACE

These Proceedings contain texts of all invited papers presented at a Symposium on Analytic Computational Complexity held by the Computer Science Department, Carnegie-Mellon University, Pittsburgh, Pennsylvania, on April 7-8, 1975. Abstracts of contributed papers are also included.

The decision to have a symposium in April, 1975 was made very informally. A number of the major international figures in analytic complexity planned to be at Carnegie-Mellon University for periods of time ranging from a month to a year. The intersection of these visits was in April. One easy way for the researchers to let each other know about their work was to have them make formal presentations. From there it was just a small step to inviting a few additional speakers and making it public. The proceedings seem a good way to show present progress and future directions in analytic complexity.

The research in the papers by R.P. Brent, B. Kacewicz, H.T. Kung, R. Meersman, J.F. Traub, and H. Woźniakowski was supported in part by the National Science Foundation under Grant GJ-32111 and the Office of Naval Research under Contract N00014-67-A-0314-0010, NR 044-422.

J.F. Traub

INTRODUCTION

J. F. Traub
Department of Computer Science
Carnegie-Mellon University
Pittsburgh, Pa.

I believe there has been more progress made in analytic computational complexity in the last two years than since the beginning of the subject around 1960. Perhaps this Symposium helped serve as a forcing function in this progress. In this introduction I would like to summarize what I believe are some of the reasons for studying complexity in general and analytic computational complexity in particular. Then I will briefly overview the invited papers which are presented in these Proceedings.

Some of the reasons for studying complexity (a partial list):

1. The selection of algorithms is a central issue in much of computer science and applied mathematics. The selection of algorithms is a multi-dimensional optimization problem. One of these dimensions is the complexity of the algorithms.

2. The literature contains countless papers giving conditions for the convergence of an infinite process. A process has to be more than convergent in order for it to be computationally interesting. We must also be able to bound (preferably a priori) its cost. One central issue of analytic computational complexity is what additional conditions must be imposed on a problem such that the cost of its solution can be a priori bound.

3. Complexity results help give structure to a field. For example we now know that the maximal order of an iterative process depends only on the information used by the iterative algorithm. We can therefore classify algorithms by the information they use.

4. Lower bounds on problem complexity give us a natural hierarchy based on the intrinsic difficulty of the problems.

5. Complexity leads to a mathematically interesting and satisfying theory. There seem to be numerous, deep questions.

I now turn to an overview of the papers presented in these proceedings.

Winograd

A general adversary principle is enunciated by Winograd and established as a primary technique for proving lower bounds. Winograd applies this principle and shows how lower bound results can be obtained in a number of problem areas.

Traub and Woźniakowski

An early and valid criticism of traditional iterative complexity theory has been that the theory is asymptotic whereas in practice only a finite and indeed often a small number of iterations are used. In this paper a non-asymptotic theory is developed with strict upper and lower bounds on complexity.

Kung

Iterative computational complexity has always been a local theory which assumes that a sufficiently good initial approximation is given and a solution is then calculated. Clearly, the right approach is: given an operator equation with certain properties, bound the complexity of finding the solution. Kung shows that if the operator satisfies conditions similar to those of the Newton-Kantorovich Theorem, then a starting approximation for Newton iteration which falls within the Kantorovich ball can be guaranteed and the complexity of the total process can be bounded.

Brent (Optimal-Order)

Algorithms for calculating zeros of a scalar function f are introduced for the case that f' is cheaper than f. The existence of algorithms of order 2ν which use one evaluation of f and ν evaluations of f' at each step is established. Meersman (these Proceedings) shows these algorithms have optimal order. Optimal non-linear Runge-Kutta methods are also defined.

Woźniakowski (Maximal Order)

Intimately connected with (although not equivalent to) the issue of minimal complexity is the problem of maximal order. Woźniakowski independently discovered the adversary principal (see discussion of Winograd's paper) and uses it as an engine for proving numerous maximal order results for operator equations.

Meersman

Using Woźniakowski's techniques for establishing maximal order, Meersman proves that the maximal order of any scalar iteration using three pieces of information is four. Furthermore, he establishes that there are exactly seven classes of information which achieve maximal order. The optimality of "Brent information" is also established.

Kacewicz

Function and derivative evaluations have traditionally been used to solve operator equations. Kacewicz shows that integrals may also be used. The general question of what information is relevant to the solution of a problem is a central one and will be an area of much further investigation.

Schultz

Results are summarized for three topics: (1) Lower and upper bounds are obtained for a generalized interpolation problem. (2) Storage and time results are obtained for a class of sparse linear problems. (3) The complexity of algorithms for solving a sparse non-linear system are analyzed.

Brent (Multiple-Precision)

Algorithms for finding high-precision approximations to simple zeros of smooth functions are analyzed. The results are applied to develop fast methods for evaluating the elementary functions, for the computation of π, and for certain functions of power series.

Woźniakowski (Stability)

An important dimension in the selection of algorithms is stability of the algorithm. Woźniakowski analyzes the stability of iterations for linear and non-linear problems and investigates the interaction between stability and complexity.

Rice

The complexity of computing approximations to real functions is surveyed and new results are reported. The approximations satisfy a variety of criteria.

Yun

The use of analytic techniques such as iteration to solve algebraic problems is a promising area. Yun shows that the Hensel Lemma and the Zassenhaus Construction may be interpreted as linearly and quadratically convergent iterations, respectively.

Brent and Kung

Techniques for doing fast multiplication and division of power series were developed some years ago. Brent and Kung develop a new upper bound for composition of power series and by using iteration show this leads to a new upper bound on the complexity of functional inversion. Since these results were obtained the week before the Symposium, only an extended abstract appears here.

SOME REMARKS ON PROOF TECHNIQUES IN ANALYTIC COMPLEXITY

S. Winograd
IBM Thomas J. Watson Research Center
Yorktown Heights, New York

Abstract

Most proofs of lower bounds in analytic complexity can be characterized as using a "fooling" technique: Given a numerical algorithm, if we find two functions which agree on the "sampled" data, then the half distance between them is a lower bound on the error of numerical method, since the numerical method cannot distinguish between these two functions.

In this paper we set up a general model for this kind of proof technique, and then use the insight gained to obtain a lower bound on the minimal error in optimal recovery of a function from its samples. Estimates of this error for some frequently occurring sampling is also given.

The main problem in analytic complexity is to find the inherent error in numerical methods. Given a function, we want to learn some of its properties, like its value at a given point, its roots, its maxima, etc. Of course, if the function were completely known, then these quantities were completely determined. In practice, however, we are given only discrete samples of the function, and therefore will be able to only approximate the desired quantities. Different numerical methods yield the results within different errors; and that raises the question of determining the "best" numerical method., i.e., the method which causes the smallest error.

Looking at many results in analytic complexity, one observes a common idea in their proofs: Given a numerical algorithm, by how much can we "fool" it? Put differently : What is the "spread" of the functions which are consistent with the given data? To show how this idea was used we will illustrate it by a few examples.

<u>Example 1</u>: Consider the set of unimodular functions on the interval $[0,1]$. The functions are not necessarily smooth, not even continuous, but to each function f we can assign a point $x \in [0,1]$ such that if $y<z<x$ then $f(y) < f(z)$ and if $x<y<z$ then $f(y) > f(z)$. We want to determine this point x by sampling the function f at points $x_1, x_2, \ldots, x_n$. The

question is what is the optimal strategy for choosing $x_1, x_2, \ldots, x_n$ so as to minimize the error in determining x. Assume $x_1, x_2, \ldots, x_n$ have been chosen, then because of the unimodularity of f we can reorder these points such that

(1) $$f(x_1) < f(x_2) < \ldots < f(x_i) > f(x_{i+1}) > \ldots > f(x_n).$$

It is obvious that x lies in the interval $\left[f(x_{i-1}), f(x_{i+1})\right]$, moreover, it can be easily verified that x can be any point in this interval, so the "best" choice of x is $(f(x_{i-1}) + f(x_{i+1}))/2$, and the error is $(f(x_{i+1}) - f(x_{i-1}))/2$. If we now choose another sample point, which will, of course, be in the interval $\left[f(x_{i-1}), f(x_{i+1})\right]$ and denote it by x_{n+1} (with no loss of generality, we will assume $x_{n+1} \in \left[f(x_{i-1}, f(x_i)\right]$, if x_{n+1} lies to the "right" of x_i the argument is symmetric). Having chosen x_{n+1} we narrowed the "spread" of the functions which are consistent with the data, and now we know that x lies either in the interval $\left[x_{i-1}, x_i\right]$ (if $f(x_{n+1}) > f(x_i)$) or in the interval $\left[x_{n+1}, x_{i+1}\right]$ (if $f(x_{n+1}) < f(x_i)$). An analysis of these situations led Kiefer [53] to the result:

<u>Theorem</u> (Kiefer): After n samples, the error ε satisfies $\varepsilon \geq \frac{1}{2u_{n+1}}$, where u_n is the nth Fibonacci number.

<u>Example 2</u>: Consider a function of a single variable with a simple root r. We want to determine the value of r by "sampling" f and some of its derivatives at the points $x_1, x_2, \ldots, x_n$ in the neighborhood of r. If we denote by r_n

the approximation to r after n samples, and by e_n the error $|r_n - r|$, then we are looking for an iterative method for which e_n converges to zero as fast as possible. The question then is to determine the fastest rate of convergence of e_n.

Assume $x_1, x_2, \ldots, x_n$ are points near the root r of a function $f(x)$, and that we have computed $f^{(j)}(x_i)$, $0 \le j \le d$, $1 \le i \le n$; how much do we know of f, and in particular, how well can we determine the root r. Brent, Winograd, and Wolfe [73] used the fact that if $f(x)$ satisfies the data, then so does $g(x) = f(x) + h(x) \prod_{i=1}^{n} (x - x_i)^{d+1}$, and therefore no algorithm can determine r with error smaller than $\frac{1}{2}|r - r'|$ (where r' is the root of $g(x)$). Using this fact they proved:

Theorem (Brent, Winograd, Wolfe): If Φ is an iterative method for finding the root of a function, and if Φ uses only $f^{(j)}(x_i)$ $0 \le j \le d$, $1 \le i \le n$ to determine the nth approximation of the root then its power of convergence cannot exceed $d+2$.

Example 3: The third example illustrates the technique of "fooling" the algorithm in numerical solution of integral equations. Consider the Fredholm equation of the second type

$$(2) \qquad f(x) = \int_D K(x,y) f(y)\, dy + h(x)$$

where D is a domain in m dimensional space and the kernel $K(x,y)$ as well as $h(x)$ have all derivatives up to the rth derivative. A numerical solution of this equation "samples"

K and h at n points, and uses this information to determine f. Let $K_1(x,y)$ and $K_2(x,y)$ be such that K_1-K_2 vanishes at these n points, then the numerical method will yield the same answer whether we want to compute

(3) $$f_1(x) = \int_D K_1(x,y) f_1(y)\,dy + h(x)$$

or

(3') $$f_2(x) = \int_D K_2(x,y) f_2(y)\,dy + h(x)$$

and therefore $\frac{1}{2}\left|f_1-f_2\right|$ is a lower bound on the inherent error of the numerical method. Using this observation Emelyanov and Ilin [67] proved:

Theorem (Emelyanov, Ilin): Any numerical method for solving (2) which uses n points can determine f only within an error proportional to $n^{-\frac{r}{2m}}$.

We will now make the reasoning of these three examples more formal. Let $U:D \to E$ be the function to be computed. In the first example D is the set of unimodular functions on $[0,1]$, E is $[0,1]$ and U is the function which determines the location of the maxima. In the second example D is the set of analytic functions on an interval J with a simple zero in J, E is J and U gives the location of the zero. In the third example, D is the space consisting of a pair $(K(x,y), h(x))$ which have up to r^{th} derivatives in a domain B, E is a set of functions on B, and U is the solution of the equation $f(x) = \int_B K(x,y) f(y)\,dy + h(x)$.

A numerical method for computing U consists of $s:D\to\bar{D}$, a "sampling" of the function in D, and then a numerical scheme $\psi:\bar{D}\to E$. The sampling function s is linear, and $\bar{D}$ is in a finite dimensional space, and therefore is, in general, many to one. Assume $\bar{d}\in\bar{D}$ be the result of the sampling, then the numerical scheme ψ has no information to differentiate between the different elements in $s^{-1}(\bar{d})$, and consequently cannot "know" which point in $Us^{-1}(\bar{d})$ to choose. Put differently, whatever point in $e = \psi(\bar{d})$ results from the numerical method, we can choose $f \in s^{-1}(\bar{d})$ such that $e' = U(f)$ is different from e, and $|e-e'|$ is a measure of the inherent error.

Let V be a set in a Banach space, then for every $v \in V$, we define $\delta(v) = \sup_{v'\in V} |v-v'|$, and the radius of V is defined by $r(V) = \inf_{v\in V} \delta(V)$. (In many cases $r(V)$ is half the diameter of V and hence the name.) It is clear that the worst case error of any numerical method cannot be smaller than $\sup_{\bar{d}\in\bar{D}} r(Us^{-1}(\bar{d}))$. Theoretically we can find a numerical method which comes arbitrarily close to the lower bound. Let $c(V)$ be a point in V such that $\delta(c(V)) \le r(V) + \varepsilon$, then choosing the numerical method ψ to be cUs^{-1} guarantees that its worst case error comes within ε of the lower bound.

In general, it is very difficult to determine the quantity $\sup_{\bar{d}\in\bar{D}} r(Us^{-1}(\bar{d}))$. However, in many situations which arise in practice the computation of $\sup_{\bar{d}\in\bar{D}} r(Us^{-1}(\bar{d}))$ can be simplified. We will assume that D is a ball in a Banach space, and that U is linear. Let $d \in s^{-1}(\bar{d})$ then $s^{-1}(\bar{d}) = (d+s^{-1}(0)) \cap D$, where 0 is the zero vector of $\bar{D}$, and therefore $Us^{-1}(\bar{d}) = (U(d) + Us^{-1}(0)) \cap E$. Consequently $r(Us^{-1}(\bar{d})) \leq r(Us^{-1}(0))$ and $\sup_{\bar{d}\in\bar{D}} r(Us^{-1}(\bar{d})) = r(Us^{-1}(0))$; that is, in trying to determine the inherent error, it is enough to focus our attention on the case that all the sampled data is 0.

The author, together with C. A. Micchelli and T. J. Rivlin tested the concepts outlined before. They considered the problem of recovery of smooth functions from its samples. Let $W_\infty^{(n)}[0,1] = \left\{f\in C^{n-1}[0,1] : f^{(n-1)} \text{abs.cont.}, f^{(n)} \in L_\infty[0,1]\right\}$. Suppose for the $n+r$ points $0 \leq x_1 \leq x_2 \leq \ldots \leq x_{n+r} \leq 1$ with $n_i < x_{n+i}$, we are given $f(x_i)$ (using the convention that if $x_{i-1} < x_i = x_{i+1} = \ldots = x_{i+s} < x_{i+s+1}$ then $f(x_{i+j}) = f^{(j)}(x_i)$ $j=0,1,\ldots,s$). If we denote by F the vector $(f(x_1), f(x_2), \ldots, f(x_{n+r}))$ then the question is how well can we recover f from F, where the goodness of the recovery is measured by the uniform norm. In other words we consider the problem where $D = \left\{f\in W_\infty^{(n)}[0,1] : |f^{(n)}| \leq 1\right\}$, U is the identity map and E is D with the uniform norm. We like to determine the quantity $E(x_1,\ldots,x_{n+r}) = \inf_{\psi} \sup_{f\in D} |f-\psi F|$. We will not describe the results in detail, and the interested reader is referred to Micchelli, Rivlin, Winograd [74].

An important technique is the perfect spline interpolation. Karlin [73] proved that, given F, there exists a perfect spline of degree n

$$(4) \qquad P_F(t) = \sum_{i=0}^{n-1} a_i x^i + d\left[t^n + 2\sum_{i=1}^{k-1} (-1)^i (t-\xi_i)_+^n\right]$$

with knots satisfying $0<\xi_1<\xi_2<\ldots<\xi_{k-1}<1$ and $k\leq r$, such that $P_F = F$. (We will denote the vector of "samples" of a function f by F.) Moreover, if $g \in W_\infty^{(n)}[0,1]$ satisfies $G = F$ then $|g^{(n)}| \geq |P_F^{(n)}|$.

Let $\xi \neq x_i$ $i=1,2,\ldots,n+r$, and consider the vector $(f(x_1),f(x_2),\ldots,f(x_{n+r}),f(\xi))$ where $f(x_i) = 0$, $i=1,2,\ldots,n+r$, and $f(\xi) = 1$. By Karlin's result there exists a perfect spline q such that $Q = F$, which Cavaretta [75] showed is unique. Define $c = \frac{|q|}{|q^{(n)}|}$ then we obtain:

<u>Theorem</u>: $c \leq E(x_1,x_2,\ldots,x_{n+r}) \leq 2c$.

The proof of the theorem follows the outline described before. Since sampling both $\frac{q}{|q^{(n)}|}$ and $\frac{-q}{|q^{(n)}|}$ yields the 0 vector, it follows that $E(x_1,\ldots,x_{n+r}) \geq \frac{1}{2}\left|\frac{q-(-q)}{|q^{(n)}|}\right| = c$, which establishes one inequality. The second inequality is proved by taking perfect spline interpolation for ψ. A more refined analysis shows that $E(x_1,\ldots,x_{n+r}) = c$.

In general it is difficult to calculate the value of e.

However, the following results can be established:

<u>Theorem</u>: 1. $c \leq \frac{1}{n!}$

2. If $x_1 = 0$, $x_{n+r} = 1$, $\Delta = \max(x_{i+1} - x_i)$ then

$$c \leq \frac{(n-1)^{n-1}}{n^n} \cdot \frac{1}{n!}$$

$$c \leq \frac{(n-1)^{n-1}}{n^n} \left(\frac{\Delta}{2}\right)^n$$

3. If $x_{jn} = x_{jn+1} = \ldots = x_{j(n+1)-1}$ then

$$c \leq \frac{n+1}{2^n \cdot n!} \left(\frac{\Delta}{2}\right)^n$$

4. $c \leq \frac{\Delta^n}{4^n \cdot n!}$

References

Brent, Winograd, Wolfe [73] Brent, R., Winograd, S., Wolfe, P., "Optimal iterative processes for root-finding," Numer. Math., Vol. 20 (1973), pp. 327-341.

Cavaretta [75] Cavaretta, A. S., "Oscillation and zero properties for perfect splines and monosplines." To appear in J. d'Analyse Math.

Emelyanov, Ilin [67] Emelyanov, K. V., and Ilin, A. M., "Number of arithmetical operations necessary for the approximate solution of Fredholm integral equation of the second kind," USSR Comput. Math. and Math. Phys., Vol. 7 (1967), pp. 259-266.

Karlin [73] Karlin, S., "Some variational problems on certain Sobolev spaces and perfect splines, " Bull. A.M.S., Vol. 79 (1973), pp. 124-128.

Kiefer [53] Kiefer, J., "Sequential minimax search for a maximum," Proceedings Amer. Math. Soc., Vol. 4, (1953), pp. 502-506.

Micchelli, Rivlin, Winograd [74] Micchelli, C. A., Rivlin, T. J. and Winograd, S., "The optimal recovery of smooth functions," IBM Research Report, RC-4920 (1974).

STRICT LOWER AND UPPER BOUNDS ON ITERATIVE COMPUTATIONAL COMPLEXITY

J. F. Traub
Department of Computer Science
Carnegie-Mellon University
Pittsburgh, Pennsylvania 15213

H. Woźniakowski
Department of Computer Science
Carnegie-Mellon University
Pittsburgh, Pennsylvania 15213
(Visiting from the University of Warsaw)

1. INTRODUCTION

Complexity is a measure of cost. The relevant costs depend on the model under analysis. The costs may be taken as units of time (in parallel computation), number of comparisons (in sorting algorithms), size of storage (in large linear systems), or number of arithmetics (in matrix multiplication). Of course a number of different costs may be relevant to a model. One can analyze the complexity of an algorithm, of a class of algorithms, or of a problem. The subject dealing with the complexity of an algorithm is usually called "Analysis of Algorithms". The subject dealing with the analysis of a class of algorithms or of a problem is called computational complexity.

Computational complexity comes in many flavors depending on the class of algorithms, the problem, and the costs. We limit ourselves here to mentioning three types of computational complexity. In each of these the costs are taken as the arithmetic operations. *Algebraic computational complexity* deals with a problem and a class of algorithms which solve the problems at finite cost. Typically the problem belongs to a class of problems which is indexed by an integer n. Let

$\text{comp}(P_n)$ be the complexity of solving the nth problem in the class. We are interested in lower bounds $L(P_n)$ and upper bounds $U(P_n)$ on $\text{comp}(P_n)$,

$$(1.1) \quad L(P_n) \le \text{comp}(P_n) \le U(P_n).$$

The upper bounds are obtained by exhibiting an algorithm for solving P_n with complexity $U(P_n)$. Lower bounds are obtained by theoretical considerations and "non-trivial" lower bounds are difficult to obtain. For example if P_n is the problem of multiplying two n by n matrices and if the cost of each arithmetic operation is taken as unity then

$$O(n^2) \le \text{comp}(P_n) \le O(n^\beta), \ \beta = \lg 7.$$

(We use lg to represent $\log_2$.) Borodin and Munro [75] survey the state of knowledge in algebraic complexity.

Exact solutions of "most" problems in science, engineering, and applied mathematics cannot be obtained with finite cost even if infinite-precision arithmetic is assumed. Indeed linear problems and evaluation of rational functions which can be solved at finite cost are the exception. Even when the problem can be solved rationally, we may choose to solve it by iteration. An example is the solution of large sparse linear systems. Typically, non-linear problems cannot be solved at finite cost.

We call the branch of complexity theory that deals with non-finite cost problems <u>analytic computational complexity</u>. Often the algorithms are iterative and we then refer to <u>iterative computational complexity</u>.

In this paper we propose a new methodology for iterative computational complexity. Our aim is to create at least a partial synthesis between iterative complexity and other types of complexity.

A basic quantity in iterative complexity has been the efficiency index of an algorithm or class of algorithms. In this paper we introduce a new quantity, the complexity index, which is the reciprocal of the efficiency index. The complexity index is directly proportional to the complexity of an algorithm or class of algorithms. We show under what conditions the complexity index is a good measure of complexity. Our methodology is non-asymptotic in the number of iterations. Earlier analyses of complexity applied only as the number of iterations went to infinity and this is not of course realistic in practice.

We summarize the contents of this paper. In Section 2 we analyze a simplified model of the errors of an iterative process and show that complexity is the product of two factors, the complexity index and the error coefficient function. Bounds on the error coefficient function are derived in the following Section and used to derive rigorous conditions for comparing the complexity of two different algorithms. In Section 4 we show how the results of the simple model can be applied to a realistic model of one-point iteration. Lower and upper bounds on the complexity index for several important classes of iterations appear in Section 5. In a short concluding Section we state the extensions and generalizations to be reported in future papers.

2. BASIC CONCEPTS

We analyze algorithms for the following problem. Let f be a non-linear real or complex scalar function with a simple zero α. Let x_0 be given and let an algorithm ϕ generate a sequence of approximations $x_1,\dots,x_k$ to α. We terminate the algorithm when x_k is a sufficiently good approximation to α. This will be made precise below.

The appropriate setting for this investigation is to consider f as a non-linear operator on a Banach space of finite or infinite dimension. Since many of the basic ideas can be illustrated when f is a non-linear scalar function we shall assume throughout this paper that this holds. We must remark however that some of the most interesting and important results deal with the dependence of complexity on problem dimension and we do not deal with that dependence here.

Let $e_i > 0$ represent some measure of the error of x_i. For example, e_i might represent

$$|x_i-\alpha|, \text{ the absolute error}$$

$$\frac{|x_i-\alpha|}{|\alpha|}, \text{ the relative error}$$

$$|f(x_i)|, \text{ the residual.}$$

Assume that the e_i satisfy the error equation

$$(2.1) \quad e_i = A_i e_{i-1}^p, \; p \geq 1, \; i = 1,2,\dots,k.$$

We call p the <u>non-asymptotic order</u> and A_i the <u>error coefficient</u>. We require $0 < L \leq A_i \leq U < \infty$ for all values of e_0 including the possibility that e_0 be arbitrarily small. Then p is unique. Many iterations satisfy the model given by (2.1). In Section 6 we mention extensions to this model.

EXAMPLE 2.1. Let the algorithm be Newton-Raphson iteration and let e_i denote the absolute error. Then

$$p = 2,\ A_i = \left|\frac{f''(\eta_i)}{2f'(x_i)}\right|,$$

where η_i is in the interval spanned by α and x_i. ■

We simplify the model of (2.1) and show what kind of results may then be obtained. In Section 4 we return to the analysis of (2.1). Let

$$(2.2)\quad e_i = Ae_{i-1}^p,\ p \geq 1,\ i = 1,\ldots,k.$$

We call this the <u>constant error coefficient model</u> while (2.1) is the <u>variable error coefficient model</u>.

We consider first the case $p > 1$. It is easy to verify that

$$(2.3)\quad e_i = e_0\left(\frac{1}{w_p}\right)^{p^i-1},\ i = 0,\ldots,k,$$

where

$$(2.4)\quad w_p = \frac{1}{A^{\frac{1}{p-1}}e_0}.$$

Choose ε', $0 < \varepsilon' < 1$, and let k be the smallest index for which $e_k \leq \varepsilon' e_0$. Define $\varepsilon \leq \varepsilon'$ so that

$$(2.5)\quad e_k = \varepsilon e_0.$$

ε' is a basic parameter which measures the increase in precision to be obtained in the iteration. We choose ε to avoid ceiling and floor functions later in this paper. It is convenient to assume $\varepsilon \leq 2^{-2}$ (we use this in Theorem 3.1) but this is non-restrictive in practice.

From (2.3), (2.5),

$$(2.6) \quad \varepsilon = \left(\frac{1}{w_p}\right)^{p^k - 1},$$

and it follows that

$$(2.7) \quad k = \frac{g(w_p)}{\lg p},$$

where

$$(2.8) \quad g(w_p) = \lg\left(1 + \frac{t}{\lg w_p}\right), \quad t = \lg\left(\frac{1}{\varepsilon}\right).$$

This is independent of the logarithm base but it is convenient to take all logarithms to base 2. Then, if e_i is the relative error, t measures the number of bits to be gained in the iteration.

We denote the complexity of iteration i by c_i. In this paper we assume $c_i = c$ is independent of i. We defer a discussion of the estimation of c until Section 5. The important case of variable cost will be considered in a future paper. We define the complexity of the algorithm by

$$(2.9) \quad \text{comp} = ck.$$

Then from (2.7), (2.8),

$$(2.10) \quad \text{comp} = zg(w_p)$$

where we define

$$(2.11) \quad z = \frac{c}{\lg p}.$$

as the <u>complexity index</u>.

We call g the <u>error coefficient function</u>. Equation (2.10) will be fundamental in our further analysis.

We have decomposed complexity into the product of two factors. The complexity index, which is independent of both the error coefficient and the starting error, is relatively easy to compute for any given algorithm. (However, lower bounds on the complexity index for classes of algorithms require upper bounds on order which is a difficult problem only solved for special cases (Kung and Traub [73], Meersman [75] and Woźniakowski [75b]).) We shall show, in a sense to be made precise in the next section, that the error coefficient is insensitive for a large portion of its domain and that complexity is determined primarily by the complexity index. We shall also show there are cases where complexity is determined primarily by the error coefficient function.

The complexity index is the reciprocal of a quantity called the efficiency index which has played an important role in iterative complexity. See, for example, Traub [64, Appendix C], Traub [72], Paterson [72] and Kung [73a]. Since complexity varies directly with the complexity index we feel that the complexity index rather than the efficiency index, should be basic.

We have been considering the case $p > 1$. For completeness we write down the case $p = 1$. Then $e_i = Ae_{i-1}$, $i = 1, 2, \ldots, k$ and $e_k = A^k e_0 = \varepsilon e_0$. Hence

$$(2.12) \quad k = \frac{t}{\lg(\frac{1}{A})}, \quad \text{comp} = \frac{ct}{\lg(\frac{1}{A})} .$$

We shall not pursue the case $p = 1$ further and shall assume for the remainder of this paper that $p > 1$, unless we state otherwise.

3. BOUNDS ON THE ERROR COEFFICIENT FUNCTION

We turn to an analysis of the error coefficient function which is one of the two factors which determines the complexity in (2.10). To see which values of w_p are of interest, note that from (2.3), $e_k < e_0$ iff $w_p > 1$. From the definition of k it is clear that $k \geq 1$ and hence from (2.7), (2.8), $w_p \leq \left(\frac{1}{\varepsilon}\right)^{\frac{1}{p-1}}$. Hence we assume

$$(3.1) \quad 1 < w_p \leq \left(\frac{1}{\varepsilon}\right)^{\frac{1}{p-1}} = 2^{\frac{t}{p-1}}.$$

Generally w_p depends on p. For many classes of iterations

$$(3.2) \quad a^{p-1} \leq A \leq b^{p-1}.$$

Then

$$\frac{1}{ae_0} \geq w_p \geq \frac{1}{be_0}$$

and the bounds on w_p are independent of p. If (3.2) holds for a class of iterations Φ we shall say that Φ is <u>normal</u>. An example of a normal class of iterations may be found in Woźniakowski [75b]. To simplify notation we shall henceforth write w_p as w whether or not we are dealing with a normal class.

Now, g(w) is a monotonically decreasing function and

$$\lim_{w \to 1^+} g(w) = \infty, \quad \lim_{w \to \infty} g(w) = 0.$$

To study the size of g(w) we somewhat arbitrarily divide the range of w, given by (3.1), into three sub-ranges.

<u>$1 < w \leq 2$</u>. Since $g(w) = \lg(t + \lg w) - \lg\lg w$ and $0 < \lg w \leq 1$, we conclude

$$\lg t - \lg\lg w < g(w) \leq \lg(1+t) - \lg\lg w.$$

$\underline{2 \le w \le t}$. Since $g(w)\downarrow$, $g(2) = \lg(1+t)$, $g(t) > \lg t - \lg\lg t$, we conclude

$$\lg t - \lg\lg t < g(w) \le \lg(1+t).$$

$\underline{t \le w \le 2^{\frac{t}{p-1}},\ 2^{\frac{t}{p-1}} \ge t}$. Then

$$\lg p \le g(w) < 1 + \lg t - \lg\lg t$$

To get some feel for the length of these sub-ranges, observe that if e_i represents relative error then in single-precision computation on a "typical" digital computer we might take $\varepsilon = 2^{-32}$. Then $t = 32$ and if $p = 2$, then $2^{\frac{t}{p-1}} = 2^{32}$.

From the bounds on the error coefficient function and (2.10) we immediately obtain the following bounds on complexity.

THEOREM 3.1. If $1 < w \le 2$,

(3.3) $z(\lg t - \lg\lg w) < \text{comp} \le z(\lg(1+t) - \lg\lg w)$.

If $2 \le w \le t$,

(3.4) $z(\lg t - \lg\lg t) \le \text{comp} \le z\lg(1+t)$.

If $t \le w \le 2^{\frac{t}{p-1}}$, (with $2^{\frac{t}{p-1}} \ge t$),

(3.5) $c \le \text{comp} < z(1 + \lg t - \lg\lg t)$. ■

We discuss some of the implications of this Theorem. As w approaches unity, then for ε fixed, comp $\sim -z\lg\lg w$. In this case the effect of the error coefficient A and the initial error e_0 cannot be neglected.

Complexity depends more on the nearness of w to unity than of ε to zero. To see this, observe that if $2 \le w \le t$, comp $\sim z\lg\lg(1/\varepsilon) = \text{comp}_1$ while if $1 < w \le 2$, comp $\sim z(\lg\lg(1/\varepsilon) - \lg\lg w) = \text{comp}_2$. Let $\varepsilon = 2^{-2^j}, w-1=2^{-2^j}\ln 2$. Then $\text{comp}_1 = jz$, $\text{comp}_2 \sim z(j+2^j)$.

Note that for any $p > 1$ the complexity of an iteration can be greater than if $p = 1$ (see (2.12)) provided w is sufficiently close to unity.

For any $w \ge 2$, complexity is bounded from above by $z\lg(1+t)$ and is therefore independent of the error coefficient A and the initial error e_0. For $w \ge 2$, complexity is insensitive to w and we need only crude bounds on w.

For $2 \le w \le t$,

$$1-\frac{\lg\lg t}{\lg t} \le \frac{\text{comp}}{z\lg t} \le 1+\frac{\lg(1+t^{-1})}{\lg t}$$

Therefore

$$1+o(1) \le \frac{\text{comp}}{z\lg t} \le 1+o(1)$$

and we conclude that on the interval [2,t] we have, for t large, very tight bounds on comp with

$$\text{(3.6)} \quad \text{comp} \sim z\lg\lg\frac{1}{\varepsilon}.$$

This should be compared with the case $p = 1$ (see (2.12)) where comp varies as $\lg\frac{1}{\varepsilon}$.

We have taken $w = 2$ as one of our endpoints for convenience but this is of course arbitrary. Any value of w sufficiently far from unity will do. If $w = 2^{\frac{1}{\nu}}$ then $g(w) = \lg(1+\nu t)$. Then the effect of the nearness of w to unity and of ε to zero are equal if $\nu = t$, that is if $w = 2^{\frac{1}{t}}$. For this choice of w, comp $= z\lg(1+t^2) \sim 2z\lg t = 2z\lg\lg\frac{1}{\varepsilon}$.

We have chosen the sub-ranges of w so that the endpoints are simple. We could also choose values of w that make the complexity formula simple. If

$$w = 2^{t/(t^u - 1)}, \quad u \geq 1, \text{ then comp} = uz\,\mathrm{lglg}\,\frac{1}{\varepsilon},$$

while if

$$w = 2^{t/(t^{\frac{1}{\nu}} - 1)}, \quad \nu \geq 1, \text{ then comp} = \frac{1}{\nu} z\,\mathrm{lglg}\,\frac{1}{\varepsilon}.$$

We now consider the methodology for <u>comparing two iterations</u> which are governed by the constant error coefficient model (2.2) and decrease the final error by the same ε. Let w_i, z_i, comp_i, $i = 1,2$ denote the parameters of the two iterations. Then

$$\frac{\mathrm{comp}_1}{\mathrm{comp}_2} = \left(\frac{z_1}{z_2}\right)\frac{g(w_1)}{g(w_2)} .$$

Clearly if $z_1 \leq z_2$ and $w_1 \geq w_2$ then $\mathrm{comp}_1 \leq \mathrm{comp}_2$. We obtain bounds on $\mathrm{comp}_1/\mathrm{comp}_2$ for sub-ranges of the w_i. Using the bounds on complexity from the previous theorem we obtain

THEOREM 3.2. If $1 < w_1, w_2 \leq 2$, then

$$\left(\frac{z_1}{z_2}\right)\left(\frac{\lg t - \mathrm{lglg}\, w_1}{\lg(1+t) - \mathrm{lglg}\, w_2}\right) < \frac{\mathrm{comp}_1}{\mathrm{comp}_2} < \left(\frac{z_1}{z_2}\right)\left(\frac{\lg(1+t) - \mathrm{lglg}\, w_1}{\lg t - \mathrm{lglg}\, w_2}\right). \tag{3.7}$$

If $1 < w_2 \leq 2 \leq w_1 \leq t$, then

$$\left(\frac{z_1}{z_2}\right)\left(\frac{\lg t - \mathrm{lglg}\, t}{\lg(1+t) - \mathrm{lglg}\, w_2}\right) \leq \frac{\mathrm{comp}_1}{\mathrm{comp}_2} < \left(\frac{z_1}{z_2}\right)\left(\frac{\lg(1+t)}{\lg t - \mathrm{lglg}\, w_2}\right). \tag{3.8}$$

If $2 \le w_1, w_2 \le t$, then

$$(3.9)\quad \left(\frac{z_1}{z_2}\right)\left(\frac{\lg t - \lg\lg t}{\lg(1+t)}\right) < \frac{\text{comp}_1}{\text{comp}_2} < \left(\frac{z_1}{z_2}\right)\left(\frac{\lg(1+t)}{\lg t - \lg\lg t}\right). \quad ■$$

We discuss some of the implications of this theorem. As $t \to \infty$, $\text{comp}_1/\text{comp}_2 \to z_1/z_2$ for any fixed values of w_1, w_2. The ratio z_1/z_2 has been the way that iterations have been compared (see Traub [64, Appendix C] where efficiency indices are used). Theorem 3.2 shows that z_1/z_2 can be a very poor measure of $\text{comp}_1/\text{comp}_2$; see for example (3.7).

Finally we observe that inequalities (3.7)-(3.9) can be rewritten to show when $\text{comp}_1 < \text{comp}_2$ or $\text{comp}_2 < \text{comp}_1$. For example, if $2 \le w_1, w_2 \le t$,

$$(3.10)\quad z_1 \le z_2\left(\frac{\lg t - \lg\lg t}{\lg(1+t)}\right), \text{ then } \text{comp}_1 < \text{comp}_2.$$

4. THE VARIABLE ERROR COEFFICIENT MODEL

We turn to the variable error coefficient model,

$$(4.1)\quad e_{i+1} = A_i e_i^p.$$

A complete analysis of this model is beyond the scope of this paper. Here we confine ourselves to the very simple assumption

$$(4.2)\quad A_L \le A_i \le A_U,\ i = 1,\ldots,k.$$

Let

$$w_L = \frac{1}{\frac{1}{A_L^{p-1} e_0}},\quad w_U = \frac{1}{\frac{1}{A_U^{p-1} e_0}}.$$

Then

(4.3) $zg(w_L) \le comp \le zg(w_U)$

Note that $w_U \le w_L$ and therefore (4.3) is compatible with g being a monotonically decreasing function. We can now draw conclusions from the constant coefficient model with A replaced by A_L or A_U.

EXAMPLE 4.1. Let α be a real zero and let J denote an interval centered at α. Assume f' does not vanish in J and let $x_0 \in J$ and such that

$$e_0 = |x_0-\alpha| \le \frac{\min_{x\in J}|f'(x)|}{\max_{x\in J}|f''(x)|} = \frac{1}{2A_U}$$

Then by Example 2.1, for Newton-Raphson iteration, $w_U \ge 2$ and <u>a priori</u>

(4.4) $comp \le c\ lg(1+t)$

The value of c is discussed in Section 5. Note that a sufficient condition for convergence is

$$e_0 < \frac{1}{A_U}$$

but with only this condition, complexity could be extremely large. ■

EXAMPLE 4.2. We seek to calculate $a^{1/2}$, that is solve $f(x) = x^2-a$. Let $a = 2^m\lambda^2$, m even, $\frac{1}{2} \le \lambda^2 < 2$. Then $a^{1/2} = 2^{m/2}\lambda$, $1/2^{1/2} \le \lambda < 2^{1/2}$. We use Newton-Raphson iteration,

$$x_{i+1} = \frac{1}{2}\left(x_i + \frac{\lambda^2}{x_i}\right).$$

Then $A_i = 1/(2x_{i-1})$. If $x_0 > \lambda$, then

$$A_L = \frac{1}{2x_0} \le A_i < \frac{1}{2\lambda} = A_U, \ i = 1,\ldots,k.$$

Hence

$$w_L = \frac{2}{1-\lambda/x_0}, \ w_U = \frac{2\lambda/x_0}{1-\lambda/x_0}.$$

Let $x_0 = 2^{1/2}$. Then $w_U \ge 2$ and comp $\le c \lg(1+t)$. To derive a lower bound on complexity one must make an assumption about the closest machine-representable number to $2^{1/2}$. We do not pursue that here. ■

5. BOUNDS ON THE COMPLEXITY INDEX

We have shown that provided w is not too close to unity, then for fixed ε, complexity depends only on the complexity index z. In this section we turn our attention to the complexity index.

Recall that $z = c/\lg p$. We begin our analysis of z by considering the cost per step, c. We distinguish between two kinds of problems.

We say a problem is explicit if the formula for f is given explicitly. For example, the calculation of $a^{1/2}$ by solving $f = x^2 - a$ is an explicit problem. The complexity of explicit problems has been studied by Paterson [72] and Kung [73a], [73b]. (Paterson and Kung take the efficiency index as basic.) We do not treat explicit problems here.

We say a problem is implicit if all we know about f are certain functionals of f. Classically the functionals are f and its derivatives evaluated at certain points. These functionals may be thought of as black boxes which deliver an output for any input. Kacewicz [75] has shown that integral functionals are of interest. The question of what functionals may be used in the solution of a problem are beyond the scope of this paper. We confine ourselves to implicit problems for the remainder of this paper.

We assume the same set of functionals is used at each step of the iteration. The set of functionals used by an iteration algorithm ϕ is called the information set $\mathfrak{N}$. Woźniakowski [75a] gives many examples of $\mathfrak{N}$. Let the <u>information complexity</u> $u = u(f,\mathfrak{N})$ be the cost of evaluating functionals on the information set $\mathfrak{N}$ and let the <u>combinatory complexity</u> $d = d(\phi)$ be the cost of combining functionals (see Kung and Traub [74b]). We assume that each arithmetic operation costs unity and denote the number of operations for one evaluation of $f^{(j)}$ by $c(f^{(j)})$. The following simple example may serve to illustrate the definition.

EXAMPLE 5.1. Let ϕ be Newton-Raphson iteration $x_{i+1} = \phi(x_i) = x_i = f(x_i)/f'(x_i)$, $i = 0,\ldots,k-1$. Then $\mathfrak{N} = \{f(x_i), f'(x_i)\}$, $u(f,\mathfrak{N}) = c(f) + c(f')$, $d(\phi) = 2$. ■

Up to this point we have illustrated the concepts with algorithms. Computational complexity deals with classes of algorithms and we turn to our central concern, lower and upper bounds on classes of algorithms. As usual the difficult problem is obtaining lower bounds. Good lower bounds may be obtained from good lower bounds on cost and good upper bounds on order. The problem of maximal order is a difficult one about which a great deal has been recently learned (Meersman [75], Woźniakowski [75a], [75b]). Part of the mathematical difficulty of the subject deals with the problem of maximal order. Note however that maximal order does not necessarily minimize complexity; we deal with this in a future paper. Upper bounds are obtained from algorithms. An interesting question here is a good upper bound on the combinatory complexity of a class of algorithms. Brent and Kung [75] have obtained a surprising new upper bound, $O(n \lg n)$, on the combinatory complexity on a family of nth order one-point iterations based on inverse interpolation.

It is convenient to index our algorithms by n, the number of elements in the information set $\mathfrak{N}$. We illustrate the issues with two examples.

EXAMPLE 5.2. Let ϕ_n denote any one-point iteration with $\mathfrak{N} = \{f(x_i), f'(x_i), \ldots, f^{(n-1)}(x_i)\}$. Let $c_f = \min_i c(f^{(i)})$. Then $u(f,\mathfrak{N}) \geq nc_f$. For simplicity we use the linear lower bound $d(\phi_n) \geq n-1$. (No non-linear lower bound is known.) A sharp upper bound on the order of one-point iteration (Traub [64], Kung and Traub [74a]) is $p \leq n$. Hence

$$z(\phi_n, f) \geq \frac{nc_f+n-1}{\lg n}$$

$$z(\phi_n, f) \geq \frac{nc_f+n-1}{\lg n} \geq \frac{3c_f+2}{\lg 3}$$

provided only that $c_f \geq 4$ (Kung and Traub [74b]). Hence for any one-point iteration with $w_L \leq t$

$$\text{(5.1)} \quad \text{comp} \geq \frac{3c_f+2}{\lg 3}(\lg t - \lg\lg t).$$

On the other hand there exists a one-point iteration which uses f, f', f'' and such that $p = 3$. Hence if $w_U \geq 2$,

(5.2) $\quad \text{comp} \le \frac{c(f)+c(f')+c(f'')}{\lg 3} \lg(1+t).$

For problems such that $c(f) \simeq c(f') \simeq c(f'') \simeq c_f$ the lower and upper bounds of (5.1) and (5.2) are close together. ■

EXAMPLE 5.3. Kung and Traub [74a] show there exists an iteration ψ_n for which the information set $\mathfrak{N}$ consists of n evaluation of f with $p(\psi_n) = 2^{n-1}$ and $d(\psi_n) = \frac{3}{2}n^2 + \frac{3}{2}n - 7$. Hence

$$z(\psi_n) = \frac{nc(f)+\frac{3}{2}n^2+\frac{3}{2}n-7}{n-1}.$$

The complexity index is minimzed (Kung and Traub [74b]) at $n^* = \text{round}[1 + (\frac{2}{3}(c(f)-4))^{1/2}] = O(c(f))^{1/2}$ and

$$z(\psi_{n^*}) = c(f) \Big/ \left(1 + \frac{\xi}{(c(f))^{1/2}}\right), \xi > 0.$$

It would only be reasonable to use this high an order iteration for very small ε. Assume $t \gg p^* = 2^{n^*-1}$.

Observe that $z(\psi_n)$ is a very "flat" function of n. Thus $z(\psi_3) = \frac{3}{2}c(f) + \frac{11}{2}$ and comparing this with $z(\psi_{n^*})$ shows we can gain only another $\frac{1}{2}c(f)$.

Let Φ denote the class of all multipoint iterations for which $w_U \ge 2$. Then

$$\text{comp}(\Phi) \le c(f)\lg(1+t) \Big/ \left(1 + \frac{\xi}{(c(f))^{1/2}}\right). \quad ■$$

We can obtain a lower bound on the complexity of the class of multipoint iterations by using an upper bound on the maximal order of any multipoint iteration and a lower bound on the combinatorial complexity. Kung and Traub [74a] conjecture that <u>any</u> iteration without memory which uses n pieces of information per step has order $p \le 2^{n-1}$. This conjecture seems difficult to prove in general (Woźniakowski [75b]) but has been established for many important cases (Kung and Traub [73], Meersman [75], Woźniakowski [75b]).

6. SUMMARY AND EXTENSIONS TO THE MODEL

We have constructed a non-asymptotic theory of iterative computational complexity with strict lower and upper bounds. In order to make the complexity ideas as accessible as possible we have limited ourselves to scalar non-linear problems. The natural setting for this work is in a Banach space of finite or infinite dimension and we shall do our analysis in this setting in a future paper. We have focussed on the simplified model $e_i = Ae_{i-1}^p$. More realistic models include some of the following features;

1. $e_i = A_i e_{i-1}^p$ under various assumptions on the structure of A_i.
2. $e_i = A_i e_{i-1}^{p_1} \cdots e_{i-m}^{p_m}$. This is the appropriate model for iterations with memory.
3. Variable cost per iteration, c_i.
4. Include round-off error. Then e_i will not converge to zero.

We plan to analyze these more realistic models in the future. We also intend to investigate additional basic properties of complexity. Our various results will be used to analyze the complexity of important problems in science and engineering.

ACKNOWLEDGMENT

We thank H. T. Kung for his comments on this paper.

REFERENCES

Borodin and Munro [75] Borodin, A. and Munro, I., The Computational Complexity of Algebraic and Numeric Problems, American Elsevier, 1975.

Brent and Kung [75] Brent, R. P. and Kung, H. T., to appear.

Kacewicz [75] Kacewicz, B., "The Use of Integrals in the Solution of Nonlinear Equations in N Dimensions," these proceedings. Also available as a CMU Computer Science Department report.

Kung [73a] Kung, H. T., "A Bound on the Multiplicative Efficiency of Iteration," Journal of Computer and System Sciences 7, 1973, 334-342.

Kung [73b] Kung, H. T., "The Computational Complexity of Algebraic Numbers," SIAM J. Numer. Anal. 12, 1975, 89-96.

Kung and Traub [74a] Kung, H. T. and Traub, J. F., "Optimal Order of One-Point and Multipoint Iteration," Journal of the Association for Computing Machinery 21, 1974, 643-651.

Kung and Traub [74b] Kung, H. T. and Traub, J. F., "Computational Complexity of One-Point and Multi-Point Iteration," Complexity of Computation, edited by R. Karp, American Mathematical Society, 1974, 149-160.

Meersman [75] Meersman, R., "Optimal Use of Information in Certain Iterative Processes," these proceedings. Also available as a CMU Computer Science Department report.

Paterson [72] Paterson, M. S., "Efficient Iterations for Algebraic Numbers," in Complexity of Computer Computations, edited by R. Miller and J. W. Thatcher, Plenum Press, New York, 1972. 41-52.

Traub [64] Traub, J. F., Iterative Methods for the Solution of Equations, Prentice-Hall, 1964.

Traub [72] Traub, J. F., "Computational Complexity of Iterative Processes," SIAM Journal of Computing 1, 1972, 167-179.

Woźniakowski [75a] Woźniakowski, H., "Generalized Information and Maximal Order of Iteration for Operator Equations," SIAM J. Numer. Anal. 12, 1975, 121-135.

Woźniakowski [75b] Woźniakowski, H., "Maximal Order of Multipoint Iterations Using n Evaluations," these proceedings. Also available as a CMU Computer Science Department report.

THE COMPLEXITY OF OBTAINING STARTING POINTS FOR SOLVING OPERATOR EQUATIONS BY NEWTON'S METHOD

H. T. Kung
Department of Computer Science
Carnegie-Mellon University
Pittsburgh, Pennsylvania

1. INTRODUCTION

Algorithms for finding a root α of a nonlinear equation $f(x) = 0$ usually consist of two phases:

1. Search phase: Search for initial approximation(s) to α.
2. Iteration phase: Perform an iteration starting from the initial approximation(s) obtained in the search phase.

Most results in analytic computational complexity assume that good initial approximations are available and deal with the iteration phase only. Since the complexity, i.e., the time, of the computation for solving $f(x) = 0$ is really the sum of the complexities of both the search and iteration phases, we propose to study both phases. Moreover, we observe that the complexities of the two phases are closely related. The speed of convergence of the iteration at the iteration phase in general depends upon the initial approximation(s) obtained in the search phase. If we spend much time in the search phase so that "good" initial approximation(s) are obtained, then we may expect to reduce the time needed in the iteration phase. This observation will be made precise in this paper. On the other hand, if we do not spend much time in the search phase and initial approximation(s) obtained are not so "good", then the complexity of the iteration phase could be extremely large, even if the corresponding iteration still converges. Some good examples of the phenomenon can be found in Traub and Wozniakowski [75]. All these show that the complexity of the iteration phase depends upon that of the search phase. Hence we feel that it is necessary to include both phases in the complexity analysis. Through this approach we can also obtain the optimal decision on when the search phase should be switched to the iteration phase, since it can be found by minimizing the total complexity of the two phases.

In this paper, we shall assume that f satisfies some property (or conditions), and include in our analysis the time needed in both the search phase and iteration phase. Note that it is necessary to assume f satisfies some property, since we have to make sure at least that there exists a root in the region to be searched. The general question we ask in the paper is how fast we can solve $f(x) = 0$ in the worst case, when f satisfies certain conditions.

In the following section we give the methodology to be used in the paper, which does not have the usual assumption that "good initial approximations are available". Instead, we assume that some property of the function f is known, i.e., f satisfies certain conditions. A useful lemma for proving lower bounds on complexity, in our methodology, is also given in the section.

Section 3 gives several relatively simple results for $f: R \to R$. The main purpose of the section is to illustrate the techniques for proving lower bounds. One of the results shows that even if we know that $M \geq f'(x) \geq m > 0$ on an interval $[a,b]$ and $f(a)f(b) < 0$, it is impossible to solve $f(x) = 0$ by a superlinearly convergent method. However, if in addition $|f''|$ is known to be bounded by a constant on $[a,b]$, then the problem can be solved superlinearly.

In Section 4 we give upper bounds on the complexity for solving certain operator equations $f(x) = 0$, where f maps from Banach spaces to Banach spaces (Theorem 4.3). This section contains the main results of the paper. A procedure (Algorithm 4.2) is given for finding points in the region of convergence of Newton's method, for f satisfying certain natural conditions. The complexity of the procedure is estimated a priori (Theorem 4.2), and the optimal branching condition on when the search phase is switched to the iteration phase is also given. We believe that the idea of the procedure can be applied to other iterative methods for f satisfying various conditions. By a preliminary version (Algorithm 4.1) of the procedure, we also establish an existence theorem (Theorem 4.1) in Section 4.

Summary and conclusions of the paper are given in the last section.

2. METHODOLOGY AND A USEFUL LEMMA FOR PROVING LOWER BOUNDS

Let φ be an algorithm for finding a root α of $f(x) = 0$ and x the approximation to α computed by φ. Denote the error of the approximation x by

$$\delta(\varphi,f) = \|x-\alpha\|,$$

where $\|\cdot\|$ is a suitable norm. Consider the problem of solving $f(x) = 0$ where the function (or operator) f satisfies some property (or some conditions). Since algorithms based on the property cannot distinguish individual functions in the class F of all functions satisfying the property, we really deal with the class F instead of individual functions in F. Define

$$\boxed{\Delta_i = \inf_{\varphi \in \Omega_i} \sup_{f \in F} \delta(\varphi, f),}$$

where Ω_i is the class of all algorithms using i units of time. Then the time t needed to approximate a root to within $\varepsilon > 0$ is the smallest i such that $\Delta_i \leq \varepsilon$, and an algorithm $\varphi \in \Omega_t$ is said to be <u>optimal for approximating a root to within $\varepsilon > 0$</u> if

$$\sup_{f \in F} \delta(\varphi, f) = \Delta_t.$$

Hence <u>the complexity of the problem is determined by the sequence $\{\Delta_i\}$</u>. We say the problem is <u>solvable</u> if $\{\Delta_i\}$ converges to zero, otherwise it is <u>unsolvable</u>. We are interested in solvable problems. For understanding the asymptotic behavior of the sequence $\{\Delta_i\}$, we study the <u>order of convergence</u> of $\{\Delta_i\}$, which is defined to be

$$p = \lim_{i \to \infty} (|\log \Delta_i|)^{1/i},$$

provided the limit exists. If $|\log \Delta_i|$ increases exponentially as $i \to \infty$, i.e., $p > 1$, we say the problem can be solved <u>superlinearly</u>. Our goal is to establish upper and lower bounds on Δ_i for given problems.

Upper bounds on Δ_i are established by algorithms. The following lemma is useful for proving lower bounds on Δ_i. The idea of the lemma has been used by many people, including

Brent, Wolfe and Winograd [73], Winograd [75], Woźniakowski [74], etc. under various settings. It is perhaps the most powerful idea so far for establishing lower bounds in analytic computational complexity.

Lemma 2.1. If for any algorithm using i units to time, there exist functions f_1, f_2 in F such that

(2.1) the algorithm cannot distinguish f_1 and f_2, and

(2.2) the minimum distance between any zero of f_1 and any zero of f_2 is $\geq 2\epsilon$,

then

$$\Delta_i \geq \epsilon.$$

Proof. Consider any algorithm using i units of time. Suppose that f_1, f_2 satisfy (2.1) and (2.2). Let α_1, α_2 be the zeros of f_1, f_2 respectively. By (2.1), the algorithm computes the same approximate x for f_1 and f_2. By (2.2)

$$|x-\alpha_1| + |x-\alpha_2| \geq |\alpha_1-\alpha_2| \geq 2\epsilon.$$

Hence either $|x-\alpha_1| \geq \epsilon$ or $|x-\alpha_2| \geq \epsilon$. ■

3. SOME RESULTS ON REAL VALUED FUNCTIONS OF ONE VARIABLE

In this section we shall give several relatively easy results to illustrate the concepts given in the preceding section, and the use of Lemma 2.1. We consider $f:[a,b]\subset R \to R$. For simplicity we assume that each function or derivative evaluation takes one unit of time and the time needed for other operations can be ignored.

Theorem 3.1. *If $f: [a,b] \to R$ satisfies the following properties*:

(3.1) *f is continuous on $[a,b]$, and*

(3.2) *$f(a) < 0$, $f(b) > 0$,*

then $\Delta_i = (b-a)/2^{i+1}$.

Proof. It is clear that by binary search we have that $\Delta_i \leq (b-a)/2^{i+1}$. Let φ be any algorithm using i evaluations. Algorithm 3.1 below constructs f_1, f_2 such that (2.1) and (2.2) hold for $\varepsilon = [(b-a)/2^{i+1}] - \delta$, and (3.1), (3.2) hold for $f = f_1$ or $f = f_2$, where $0 < \delta < [(b-a)/2^{i+1}]$. We first define

$$u(x) = 1,$$
$$v(x) = -1$$

for $x \in [a,b]$, and assume the first evaluation is at x_0.

Algorithm 3.1.

1. Set $\ell \leftarrow a$, $r \leftarrow b$, $m \leftarrow x_0$, $c(x) = u(x)$ for $x \in (a,b]$ and $c(a) = v(a)$.

2. If $m \notin [\ell,r]$, go to step 4.

3. If $m-\ell \geq r-m$, set $r \leftarrow m$. Otherwise, define $c(x) = v(x)$ for $x \in [\ell,m]$ and set $\ell \leftarrow m$.

4. Apply algorithm φ to function $c(x)$ and compute the next approximation.

5. If algorithm φ has not terminated, set m to be the point where the next evaluation takes place and go to step 2.

6. Define f_1, f_2 by $f_1(x) = f_2(x) = c(x)$ for $x \in [a,\ell] \cup [r,b]$,

$$f_1(x) = \begin{cases} u(x) & \text{for } x \in [\ell+\delta, r), \\ \frac{x-\ell}{\delta}[u(\ell+\delta) - v(\ell)] + v(\ell) & \text{for } x \in (\ell, \ell+\delta), \end{cases}$$

and

$$f_2(x) = \begin{cases} v(x) & \text{for } x \in (\ell, r-\delta], \\ \frac{x-r}{\delta}[u(r) - v(r-\delta)] + u(r) & \text{for } x \in (r-\delta, r). \end{cases}$$

It is straightforward to check that $r-\ell \geq (b-a)/2^i$ and that the distance between any zero of f_1 and any zero of f_2 is $\geq r-\ell - 2\delta$. Hence (2.2) is satisfied for $\varepsilon = [(b-a)/2^{i+1}] - \delta$. It is also easy to see that (2.1), (3.1) and (3.2) hold for f_1, f_2. Hence by Lemma 2.1, we have $\Delta_i \geq [(b-a)/2^{i+1}] - \delta$. Since δ can be chosen arbitrarily small, we have shown $\Delta_i \geq (b-a)/2^{i+1}$. ■

Theorem 3.1 establishes that binary search is optimal for finding a zero of f satisfying (3.1) and (3.2). The result is well-known. The theorem is included here because its proof is instructive. By slightly modifying the proof of Theorem 3.1, we obtain the following result:

Theorem 3.2. If $f: [a,b] \to R$ satisfies the following properties:

(3.3) $f'(x) \geq m > 0$ for all $x \in [a,b]$, and
$f(a) < 0$, $f(b) > 0$,

then $\Delta_i = (b-a)/2^{i+1}$.

Proof. The proof is the same as that of Theorem 3.1, except that the functions u, v are now defined as

$$u(x) = m(x-a),$$
$$v(x) = m(x-b),$$

and the functions f_1, f_2 have to be smoothed so that they satisfy (3.3). ■

One can similarly prove the following two theorems.

Theorem 3.3. If $f: [a,b] \to R$ satisfies the following properties:

$$f'(x) \leq M \text{ for all } x \in [a,b], \text{ and}$$
$$f(a) < 0,\ f(b) > 0,$$

then $\Delta_i = (b-a)/2^{i+1}$.

By Theorems 3.2 and 3.3, we know that even if f' is bounded above or bounded below, we still cannot do better than binary search in the worst case sense.

Theorem 3.4. If $f: [a,b] \to R$ satisfies the following properties:

$$M \geq f'(x) \geq m > 0 \text{ for all } x \in [a,b], \text{ and}$$
$$f(a) < 0,\ f(b) > 0$$

then $\Delta_i \geq (b-a)[(1-\frac{m}{M})^2/2]^{i+1}$.

Under the conditions of Theorem 3.4, Micchelli and Miranker [75] showed that

$$\Delta_i \leq \frac{1}{2}(b-a)(1-\frac{m}{M})^{\frac{i}{2}}.$$

Hence their algorithm is better than binary search when $\frac{m}{M} \geq \frac{3}{4}$. However, by Theorem 3.4, we know that the problem cannot be solved superlinearly, even when f' is known to be bounded above and below by some constants. In order to assure that the problem can be solved superlinearly we have to make further assumptions on the function f. A natural way is to assume that $|f''|$ is bounded. This leads to the following

Theorem 3.5. If the conditions of Theorem 3.4 are satisfied and $|f''| \leq K$ on $[a,b]$, then the problem of finding a root of $f(x) = 0$ can be solved superlinearly.

Proof. We can use binary search to find a point x_0 which satisfies the conditions of the Newton-Kantorovich Theorem (see the next section for the statement of the theorem). It is easy to see that only a finitely many steps of binary search are needed to find x_0. Starting from x_0 the Newton iterates converge to a root superlinearly. ■

It should be noted that the binary search used in the above proof would not make sense for operators mapping from Banach spaces to Banach spaces. In the following section we propose a general technique for obtaining starting points for the solution of operator equations.

4. A PROCEDURE TO OBTAIN GOOD STARTING POINTS FOR NEWTON'S METHOD

In this section we consider $f: D \subset B_1 \to B_2$, where B_1 and B_2 are Banach spaces and assume that f is Fréchet differentiable. We shall give a procedure to obtain a point x_0 such that Newton's method starting from x_0 will converge to a root α of $f(x) = 0$, provided that f satisfies some natural conditions. The use of Newton's method is only illustrative. The principle of the procedure can be applied to other iterative methods.

Let $S_r(x_0)$ denote a ball in D with center x_0 and radius r. Sufficient conditions for the quadratic convergence of Newton's method, starting from x_0, are given by the famous Newton-Kantorovich Theorem (see, e.g., Ortega and Rheinboldt [70, Section 12.6.2]), which essentially states the following:

If

(4.1) $[f'(x_0)]^{-1}$ exists, $\|[f'(x_0)]^{-1}\| \leq \beta_0$,

(4.2) $\|[f'(x_0)]^{-1} f(x_0)\| \leq \xi_0$,

(4.3) $\|f'(x)-f'(y)\| \leq K\|x-y\|$, $x,y \in S_r(x_0)$

and if

(4.4) $h_0 \equiv \beta_0 K \xi_0 < \frac{1}{2}$,

(4.5) $2\xi_0 \leq r$

then Newton's method, starting from x_0, will generate iterates converging quadratically to a root α of $f(x) = 0$ and

(4.6) $\|x_0-\alpha\| \leq 2\xi_0$.

For convenience, we say x_0 is a <u>good starting point</u> for approximating α by Newton's method or a good starting point for short, if conditions (4.1) ~ (4.5) are satsified. Note that the existence of a good starting point implies the existence of a zero of f in $S_q(x_0)$, $q = 2\xi_0$. The conditions for a point to be a good starting point are quite restrictive for certain applications. We are interested in designing algorithms for finding good starting points under relaxed conditions. We shall first prove the following existence theorem by combining the ideas of the Newton-Kantorovich Theorem and the continuation method (see, e.g. Ortega and Rheinboldt [70,

Section 7.5]). The algorithm (Algorithm 4.1) used in the proof is then developed to be a procedure (Algorithm 4.2) to obtain good starting points, for f satisfying some natural conditions.

<u>Theorem 4.1.</u> <u>If f' satisfies a Lipschitz condition on $S_{2r}(x_0) \subseteq D$,</u>

$$\|f(x_0)\| \leq \eta_0,$$

$$\|[f'(x)]^{-1}\| \leq \beta \text{ for all } x \in S_r(x_0), \text{ and}$$

$$\beta\eta_0 < r/2, \tag{4.7}$$

<u>then there exists a root of $f(x) = 0$ in $S_r(x_0)$.</u>

<u>Proof.</u> We assume that

$$\|f'(x)-f'(y)\| \leq K\|x-y\|, \quad x, y \in S_{2r}(x_0).$$

The proof is based on the following algorithm.

<u>Algorithm 4.1.</u>

The algorithm takes f satisfying the conditions of Theorem 4.1 as input and produces a good starting point for approximating a root α of $f(x) = 0$.

1. Set $h_0 \leftarrow \beta^2 K\eta_0$ and $i \leftarrow 0$.
   Pick any number δ in $(0,\frac{1}{2})$.

2. If $h_i < \frac{1}{2}$, x_i is a good starting point for approximating α and Algorithm 4.1 terminates.

3. Set $\lambda_i \leftarrow (\frac{1}{2}-\delta)/h_i$, and

$$f_i(x) \leftarrow [f(x)-f(x_i)] + \lambda_i f(x_i). \tag{4.8}$$

4. (It will be shown later that x_i is a good starting point for approximating a zero, denoted by x_{i+1}, of f_i.) Apply Newton's method to f_i, starting from x_i, to find x_{i+1}.

5. (Assume that the exact x_{i+1} is found.) Set $\eta_{i+1} \leftarrow \|f(x_{i+1})\|$ and $h_{i+1} \leftarrow \beta^2 K \eta_{i+1}$.

6. Set $i \leftarrow i+1$ and return to step 2.

In the following we prove the correctness of the algorithm. First we note that $\lambda_i \in (0,1)$ and by (4.8)

(4.9) $\eta_{i+1} = (1-\lambda_i)\eta_i$.

We shall prove by induction that

(4.10) $\|x_i - x_{i-1}\| \le 2\beta\lambda_{i-1}\eta_{i-1}$, and

(4.11) $\|x_i - x_0\| \le r$.

They trivially hold for $i = 0$.

Suppose that (4.10) and (4.11) hold and $h_i \ge \frac{1}{2}$. By (4.8),

(4.12) $\beta^2 K \|f_i(x_i)\| \le \beta^2 K \lambda_i \eta_i = \lambda_i h_i = \frac{1}{2} - \delta$,

and by (4.9),

$$2\beta\|f_i(x_i)\| \le 2\beta\lambda_i\eta_i < 2\beta\eta_i \le 2\beta\eta_0 < r.$$

Further, by (4.11), we have $S_r(x_i) \subseteq S_{2r}(x_0)$. Hence x_i is a good starting point for approximating the zero x_{i+1} of f_i. From (4.6), we know

(4.13) $\|x_{i+1} - x_i\| \le 2\beta\lambda_i\eta_i$.

Hence (4.10) holds with i replaced by i+1. By (4.13), (4.9) and (4.7), we have

$$(4.14)\quad \|x_{i+1}-x_0\| \le \|x_{i+1}-x_i\| + \|x_i-x_{i-1}\| + \dots + \|x_1-x_0\|$$
$$\le 2\beta(\lambda_i\eta_i + \lambda_{i-1}\eta_{i-1} + \dots + \lambda_0\eta_0)$$
$$\le 2\beta((1-\lambda_{i-1})\eta_{i-1} + \lambda_{i-1}\eta_{i-1} + \dots + \lambda_0\eta_0)$$
$$= 2\beta(\eta_{i-1} + \lambda_{i-2}\eta_{i-2} + \dots + \lambda_0\eta_0)$$
$$\le \dots$$
$$\le 2\beta\eta_0 < r,$$

i.e., (4.11) holds with i replaced by i+1.

We now assume that (4.10) and (4.11) hold and $h_i < \frac{1}{2}$. By (4.7) and (4.9), $2\beta\|f(x_i)\| = 2\beta\eta_i < 2\beta\eta_0 < r$. Further by (4.11), $S_r(x_i) \subseteq S_{2r}(x_0)$. Hence x_i is a good starting point for approximating α.

It remains to show that the loop starting from step 2 is finite. Suppose that $h_0 \ge \frac{1}{2}$. Since $\lambda_i \in (0,1)$ for all i, we have

$$\lambda_i = \frac{\frac{1}{2}-\delta}{\beta^2K\eta_i} = \frac{\frac{1}{2}-\delta}{\beta^2K(1-\lambda_{i-1})\eta_{i-1}} > \frac{\frac{1}{2}-\delta}{\beta^2K\eta_{i-1}} = \lambda_{i-1}, \text{ for all i.}$$

Hence by (4.9) $\eta_{i+1} = (1-\lambda_i)\eta_i < (1-\lambda_0)\eta_i$
$$< \dots$$
$$< (1-\lambda_0)^{i+1}\eta_0.$$

This implies that $h_i < \frac{1}{2}$ when $\beta^2K(1-\lambda_0)^i\eta_0 < \frac{1}{2}$, i.e., when

$$(4.15)\quad (1-\lambda_0)^i < \frac{1}{2h_0}.$$

Since $1-\lambda_0 < 1$, (4.15) is satisfied for large i. Therefore when i is large enough, $h_i < \frac{1}{2}$ and hence Algorithm 4.1 terminates. The proof of Theorem 4.1 is complete. ■

Note that Theorem 4.1 is trivial for the scalar case (i.e., when $f: R \to R$), since the mean value theorem can be used there. The problem becomes nontrivial for nonscalar cases. The main reason for including Theorem 4.1 here is to introduce Algorithm 4.1, which works for Banach spaces. It should be noted that some assumptions (e.g., (4.7)) of the theorem could be weakened by complicating the algorithm used in the proof. Theorem 4.1 is similar to a result of Avila [70, Theorem 4.3], where, instead of the assumption (4.7) used here, a more complicated assumption involving β, k, η_0 was used. Also the idea of his algorithm is basically different from that of Algorithm 4.1.

An upper bound $N(h_0, \delta)$ on the number of times the loop starting from step 2 is executed in Algorithm 4.1 can be obtained from (4.15). Since $\lambda_0 = (\frac{1}{2}-\delta)/h_0$, (4.15) is equivalent to

$$(4.16) \quad \left(1 - \frac{\frac{1}{2}-\delta}{h_0}\right)^i \leq \frac{1}{2h_0},$$

from which $N(h_0, \delta)$ can be calculated. Asymptotically, we have

$$(4.17) \quad N(h_0, \delta) \sim \frac{2h_0 \ln h_0}{1-2\delta}, \text{ as } h_0 \to \infty.$$

From (4.16) and (4.17) it appears that we should use small δ in the algorithm. However, small δ tends to slow down the convergence of the Newton iterates in step 4 of the algorithm (see (4.18), (4.19)). The problem of how to choose a suitable δ will be further discussed after Algorithm 4.2.

In Algorithm 4.1, we assume that the exact zero x_{i+1} of f_i can be found by Newton's method. This is clearly not the case in practice. Fortunately, this problem can be solved by modifying Algorithm 4.1. The modified algorithm, Algorithm 4.2, appears in the proof of the following theorem. In the theorem and the rest of the paper, a Newton step means the computation of $x - [f'(x)]^{-1}f(x)$, given x, f and f'. Hence a Newton step involves one evaluation of f, one of f' at x and the computation of $x - [f'(x)]^{-1}f(x)$ from x, f(x), f'(x).

Theorem 4.2. Suppose that the conditions of Theorem 4.1 are satisfied and

$$\|f'(x)-f'(y)\| \le K\|x-y\|, \quad x,y \in S_{2r}(x_0).$$

Then a good starting point for approximating the root of $f(x) = 0$ by Newton's method can be obtained in $N(\delta)$ Newton steps, where δ is any number in $(0,\frac{1}{2})$ and $N(\delta)$ is defined as follows. If $h_0 = \beta^2 K\eta_0 \le \frac{1}{2}-\delta$ then $N(\delta) = 0$ else $N(\delta)=I(\delta)J(\delta)$, where $I(\delta)$ is the smallest integer i such that

$$\left(1-\frac{\frac{1}{2}-\delta}{h_0}\right)^i \le \left(\frac{1}{2}-\delta\right)\frac{1}{h_0},$$

and $J(\delta)$ is the smallest integer j such that

$$(4.18) \quad 2^{\frac{1}{j-1}}(1-2\delta)^{2^j-1}(m+\beta\eta_0) \le r - 2\beta\eta_0,$$

$$(4.19) \quad \frac{1}{2^j}(1-2\delta)^{2^j-1}(m+\beta\eta_0) \le m,$$

with $m = \min(\frac{r}{2}-\beta\eta_0, \frac{\delta}{2\beta K})$.

Proof. The proof is based on the following algorithm, which

is adapted from Algorithm 4.1.

<u>Algorithm 4.2</u>.

The algorithm takes f satisfying the conditions of Theorem 4.2 as input and produces a good starting point for approximating a root α of $f(x) = 0$.

1. Set $h_0 \leftarrow \beta^2 K\eta_0$, $\bar{x}_0 \leftarrow x_0$ and $i \leftarrow 0$.
   Pick any number δ in $(0,\frac{1}{2})$.

2. If $h_i \le \frac{1}{2}-\delta$, $\bar{x}_i$ is a good starting point for approximating α and Algorithm 4.2 terminates.

3. Set $\lambda_i \leftarrow (\frac{1}{2}-\delta)/h_i$,

$$f_i(x) \leftarrow [f(x)-\eta_i f(x_0)/\eta_0] + \lambda_i \eta_i f(x_0)/\eta_0, \text{ and}$$

$$\eta_{i+1} \leftarrow (1-\lambda_i)\eta_i. \tag{4.20}$$

4. Apply Newton's method to f_i, starting from $\bar{x}_i$, to find an approximation $\bar{x}_{i+1}$ to a zero x_{i+1} of f_i such that

$$\|\bar{x}_{i+1}-x_{i+1}\| \le r - 2\beta\eta_0, \text{ and} \tag{4.21}$$

$$\|[f_i'(\bar{x}_{i+1})]^{-1} f_i(\bar{x}_{i+1})\| \le \min(\frac{r}{2}-\beta\eta_{i+1}, \frac{\delta}{2\beta K}) \tag{4.22}$$

5. Set $h_{i+1} \leftarrow \beta^2 K\eta_{i+1}$.

6. Set $i \leftarrow i+1$ and return to step 2.

Note that the h_i, λ_i, η_i, f_i, x_i in Algorithm 4.2 are the same h_i, λ_i, η_i, f_i, x_i in Algorithm 4.1. Note also that by (4.21) and (4.14) we have

$$\|\bar{x}_i-x_0\| \le \|\bar{x}_i-x_i\| + \|x_i-x_0\| \tag{4.23}$$

$$\leq (r - 2\beta\eta_0) + 2\beta\eta_0 = r, \forall i.$$

It is clear that if $h_0 < \frac{1}{2} - \frac{\delta}{2}$, $\bar{x}_0$ is a good starting point for approximating α. Now suppose $h_0 > \frac{1}{2} - \frac{\delta}{2}$. Since $\bar{x}_0 = x_0$, in the proof of Theorem 4.1 we have shown that $\bar{x}_0$ is a good starting point for approximating x_1, a zero of f_0. Let z_j denote the jth Newton iterate starting from $\bar{x}_0$ for approximating x_1. Since

$$\beta^2 K \|f_0(\bar{x}_0)\| = \beta^2 K \lambda_1 \eta_0 = \frac{1}{2} - \delta,$$

it is known (see e.g. Rall [71, Section 22]) that

$$\|z_j - x_1\| \leq \frac{1}{2^{j-1}}(1-2\delta)^{2^j-1} \|[f'(x_0)]^{-1} f(x_0)\| \text{ and}$$

$$\|[f_0'(z_j)]^{-1} f_0(z_j)\| \leq \frac{1}{2^j}(1-2\delta)^{2^j-1} \|[f'(x_0)]^{-1} f(x_0)\|.$$

Hence we may let $\bar{x}_1$ be z_j for j large enough, say, $j=j(\delta)$, then

$$\|\bar{x}_1 - x_1\| \leq r - 2\beta\eta_0, \text{ and}$$

$$\|[f_0'(\bar{x}_1)]^{-1} f_0(\bar{x}_1)\| \leq \min\left(\frac{r}{2} - \beta\eta_1, \frac{\delta}{2\beta K}\right),$$

i.e., (4.21) and (4.22) hold for $i = 0$.

Suppose that (4.21) and (4.22) hold. Then

$$(4.24) \quad \|[f'(\bar{x}_{i+1})]^{-1} f(\bar{x}_{i+1})\|$$

$$\leq \|[f_i'(\bar{x}_{i+1})]^{-1} f_i(\bar{x}_{i+1})\| + \|[f_i'(\bar{x}_{i+1})]^{-1} [f(\bar{x}_{i+1}) - f_i(\bar{x}_{i+1})]\|.$$

$$\leq \min\left(\frac{r}{2} - \beta\eta_{i+1}, \frac{\delta}{2\beta K}\right) + \beta\eta_{i+1}, \text{ and}$$

$$(4.25) \quad \|[f_{i+1}'(\bar{x}_{i+1})]^{-1} f_{i+1}(\bar{x}_{i+1})\|$$

$$\leq \|[f'(\bar{x}_{i+1})]^{-1}[f(\bar{x}_{i+1}) - \eta_{i+1}f(x_0)/\eta_0]\|$$

$$+ \|[f'(\bar{x}_{i+1})]^{-1}\lambda_{i+1}\eta_{i+1}f(x_0)/\eta_0]\|$$

$$\leq \|[f_i'(\bar{x}_{i+1})]^{-1}f_i(\bar{x}_{i+1})\| + \lambda_{i+1}\beta\eta_{i+1}.$$

Suppose that $h_{i+1} < \frac{1}{2} - \delta$. We want to prove that $\bar{x}_{i+1}$ is a good starting point for approximating α. By (4.24),

$$\beta K\|[f'(\bar{x}_{i+1})]^{-1}f(\bar{x}_{i+1})\|$$

$$\leq \beta K \cdot \frac{\delta}{2\beta K} + h_{i+1} < \frac{\delta}{2} + \frac{1}{2} - \delta = \frac{1}{2} - \frac{\delta}{2}.$$

Let $a = \|f'(\bar{x}_{i+1})]^{-1}f(\bar{x}_{i+1})\|$. If $x \in S_{2a}(\bar{x}_{i+1})$, then

$$\text{(4.26)} \quad \|x-x_0\| \leq \|x-\bar{x}_{i+1}\| + \|\bar{x}_{i+1}-x_0\|$$

$$\leq 2a + r$$

$$\leq 2(\frac{r}{2} - \beta\eta_{i+1} + \beta\eta_{i+1}) + r = 2r,$$

i.e., $x \in S_{2r}(x_0)$. Hence $\bar{x}_{i+1}$ is a good starting point for approximating α. We now assume that $h_{i+1} > \frac{1}{2} - \delta$, and want to prove that $\bar{x}_{i+1}$ is a good starting point for approximating x_{i+2}, a zero of f_{i+1}.

We have by (4.25) and (4.22),

$$\beta K\|[f_{i+1}'(\bar{x}_{i+1})]^{-1}f_{i+1}(\bar{x}_{i+1})\|$$

$$\leq \frac{\delta}{2} + \lambda_{i+1}\beta^2 K\eta_{i+1}$$

$$= \frac{\delta}{2} + \frac{1}{2} - \delta = \frac{1}{2} - \frac{\delta}{2}.$$

Let $b = \|[f'_{i+1}(\bar{x}_{i+1})]^{-1}f_{i+1}(\bar{x}_{i+1})\|$. If $x \in S_{2b}(\bar{x}_{i+1})$, as in (4.26) we can prove that $x \in S_{2r}(x_0)$. Hence $\bar{x}_{i+1}$ is a good starting point for approximating x_{i+2}. By the same argument as used for obtaining $\bar{x}_0$ and by (4.25), one can prove that if $\bar{x}_{i+2}$ is set to be the $J(\delta)$-th Newton iterate starting from $\bar{x}_{i+1}$, then

$$\|\bar{x}_{i+2}-x_{i+1}\| \le r - 2\beta\eta_0, \text{ and}$$

$$\|[f'_{i+1}(\bar{x}_{i+2})]^{-1}f_{i+1}(\bar{x}_{i+2})\| \le \min(\frac{r}{2} - \beta\eta_{i+2}, \frac{\delta}{2\beta K}),$$

i.e. (4.21) and (4.22) hold with i replaced by i+1. This shows that we need to perform at most $J(\delta)$ Newton steps at step 4 of Algorithm 4.2 to obtain each $\bar{x}_{i+1}$. Furthermore, from an inequality similar to (4.16) it is easy to see that the loop starting from step 2 of Algorithm 4.2 is executed at most $I(\delta)$ times. Therefore, for any $\delta \in (0,\frac{1}{2})$, to obtain a good starting point we need to perform at most $N(\delta)=I(\delta)\cdot J(\delta)$ Newton steps. The proof of Theorem 4.2 is complete. ■

We have shown that Algorithm 4.2 with parameter $\delta \in (0,\frac{1}{2})$ finds a starting point for Newton's method, with respect to f satisfying the conditions of Theorem 4.2, in $N(\delta)$ Newton steps. One should note that δ should not be chosen to minimize the complexity of Algorithm 4.2. Instead, δ should be chosen to minimize the complexity of the corresponding algorithm for finding root α of the equation $f(x) = 0$, which is defined as follows:

1. Search phase: Perform Algorithm 4.2.
2. Iteration phase: Perform Newton's method starting from the point obtained by Algorithm 4.2.

Note that the choice of δ determines the terminating condition of Algorithm 4.2 and hence determines when the search phase is switched to the iteration phase. Therefore the optimal time to switch can be obtained by choosing δ to minimize the sum of the complexities of the two phases.

An upper bound on the complexity of the search phase is the time needed for performing $N(\delta)$ Newton steps. Suppose that we went to approximate α to within ϵ, for a given $\epsilon > 0$. It can be shown that an upper bound on the complexity of the iteration phase is the time needed for performing $T(\delta, \epsilon)$ Newton steps, where $T(\delta,\epsilon)$ is the smallest integer k such that

$$(4.27) \quad \frac{1}{2^{k-1}}(1-2\delta)^{2^k-1}(m+\beta\eta_0) \le \epsilon,$$

(see (4.18)). Therefore, we have proved the following result.

Theorem 4.3. *If f satisfies the conditions of Theorem 4.2, then the time needed to locate a root of f(x) within a ball of radius ϵ is bounded above by the time needed to perform $R(\epsilon)$ Newton steps, where $R(\epsilon) = \min_{0<\delta<\frac{1}{2}} (N(\delta)+T(\delta,\epsilon))$, $N(\delta)$ is defined in Theorem 4.2 and $T(\delta,\epsilon)$ is defined by (4.27).*

For large h_0, we know from (4.17) that $R(\epsilon)$ grows like $O(h_0 \ln h_0)$ as $h_0 \to \infty$. For fixed h_0, we can calculate the values of $R(\epsilon)$ numerically. We have computed the values of $R(\epsilon)$ for f satisfying the conditions of Theorem 4.3 with

1. $\beta\eta_0 \le \frac{4}{5}\cdot\frac{r}{2}$,
2. $1 \le h_0 \equiv \beta^2 K\eta_0 \le 10$,

and for ϵ equal to $10^{-i}r$, $1 \le i \le 10$. Table 1 reports the results for ϵ equal to $10^{-6}r$.

h_0	δ_0	$I(\delta_0)$	$J(\delta_0)$	$N(\delta_0)$	$T(\delta_0,\varepsilon)$	$R(\varepsilon)$
1	.165	3	2	6	5	11
2	.103	8	3	24	6	30
3	.118	16	3	48	6	54
4	.129	25	3	75	6	81
5	.137	35	3	105	6	111
6	.144	47	3	141	5	146
7	.149	59	3	177	5	182
8	.154	72	3	216	5	221
9	.159	85	3	255	5	260
10	.163	99	3	297	5	302

TABLE 1

In the table, δ_0 is the δ in $(0,\frac{1}{2})$ which minimizes $N(\delta) + T(\delta,\varepsilon)$, i.e., $R(\varepsilon) = T(\delta_0) + T(\delta_0,\varepsilon)$. Suppose, for example, that $h_0 = 9$ or $h_0 \leq 9$. Then by Algorithm 4.2 with $\delta = .159$ and by (4.27), we know that the search phase can be done in 255 Newton steps and the iteration phase in 5 Newton steps. Hence a root can be located within a ball of radius $10^{-6}r$ by 260 Newton steps.

5. SUMMARY AND CONCLUSIONS

The search and iteration phases should be studied together. A methodology for studying the worst case complexity of the two phases is proposed. Results based on the methodology are <u>global</u> and <u>non-asymptotic</u> (see Theorems 4.2 and 4.3), while the usual results in analytic computational complexity are local and asymptotic. Optimal time for switching from the search phase to the iteration phase can also be determined from the methodology. A useful lemma for proving lower bounds on complexity is identified. Several optimality results are obtained for scalar functions.

The main results of the paper deal with the complexity of solving certain nonlinear operator equations f(x) = 0. Upper bounds are established by a new procedure for obtaining starting points for Newton's method. The procedure finds points where the conditions of the Newton-Kantorovich Theorem are satisfied. It is believed that the principle of the procedure can be used for other iterative schemes, where Newton-Kantorovich-like theorems are available, for f satisfying various kinds of conditions. It is not known, however, at this moment whether or not the number of Newton steps used by the procedure is close to the minimum. The problem of establishing lower bounds on the complexity for solving f(x) = 0 with f satisfying the conditions such as that of Theorem 4.3 deserves further research. The lemma mentioned above may be useful for the problem.

We end the paper by proposing an open question: Suppose that the conditions of the Newton-Kantorovich Theorem hold. Is Newton's method optimal or close to optimal, in terms of the numbers of function and derivative evolutions required to approximate the root to within a given tolerance?

ACKNOWLEDGMENTS

The author would like to thank J. F. Traub for his comments and G. Baudet, D. Heller, A. Werschulz for checking a draft of this paper.

REFERENCES

Avila [70] Avila, J., "Continuation Methods for Nonlinear Equations," Technical Report TR-142, Computer Science Center, University of Maryland, January, 1971.

Brent, Wolfe and Winograd [73] Brent, R. P., Winograd, S. and Wolfe, P., "Optimal Iterative Processes for Root-Finding, _Numer. Math._ 20 (1973), 327-341.

Micchelli and Miranker [75] Micchelli, C. A. and Miranker, W. L., "High Order Search Methods for Finding Roots," _J.ACM_ 22 (1975), 51-60.

Ortega and Rheinboldt [70] Ortega, J. M. and Rheinboldt, W. C., _Iterative Solution of Nonlinear Equations in Several Variables_, Academic Press, New York, 1970.

Rall [69] Rall, L. B., _Computational Solution of Nonlinear Operator Equations_, John Wiley and Sons, Inc., New York, 1969.

Traub and Woźniakowski [75] Traub, J. F. and Woźniakowski, H., "Strict Lower and Upper Bounds on Iterative Computational Complexity," these Proceedings.

Winograd [75] Winograd, S., "Optimal Approximation from Discrete Samples," these Proceedings.

Woźniakowski [74] Woźniakowski, H., "Maximal Stationary Iterative Methods for the Solution of Operator Equations," _SIAM J. Numer. Anal._ 11 (1974), 934-949.

J. Moré has pointed out to the author that Theorem 4.1 was previously established by A. Ostrowski using a completely different approach.

A CLASS OF OPTIMAL-ORDER ZERO-FINDING METHODS USING DERIVATIVE EVALUATIONS

Richard P. Brent

Computer Centre,
Australian National University,
Canberra, A.C.T. 2600, Australia

1. INTRODUCTION

It is often necessary to find an approximation to a simple zero ζ of a function f, using evaluations of f and f'. In this paper we consider some methods which are efficient if f' is easier to evaluate than f. Examples of such functions are given in Sections 5 and 6.

The methods considered are stationary, multipoint, iterative methods, "without memory" in the sense of Traub [64]. Thus, it is sufficient to describe how a new approximation (x_1) is obtained from an old approximation (x_0) to ζ. Since we are interested in the order of convergence of different methods, we assume that f is sufficiently smooth near ζ, and that x_0 is sufficiently close to ζ. Our main result is:

<u>Theorem 1.1</u>

There exist methods, of order 2ν, which use one evaluation of f and ν evaluations of f' for each iteration.

By a result of Meersman and Wozniakowski, the order 2ν is the highest possible for a wide class of methods using the same information (i.e., the same number of evaluations of f and f' per iteration): see Meersman [75]. The "obvious" interpolatory methods have order $\nu + 1$, but the optimal order 2ν may be obtained by evaluating f' at the correct points. These points are determined by some properties of orthogonal and "almost orthogonal" polynomials.

If $\nu + 1$ evaluations of f are used, instead of one function evaluation and ν derivative evaluations, then the optimal order is 2^ν for methods without memory (Kung and Traub [73,74], Wozniakowski [75a,b]), and $2^{\nu+1}$ for methods with memory (Brent, Winograd and Wolfe [73]). Thus, our methods are only likely to be useful for small ν or if f' is much cheaper than f.

Special Cases

Our methods for $\nu \geq 3$ appear to be new. The cases $\nu = 1$ (Newton's method) and $\nu = 2$ (a fourth-order method of Jarratt [69]) are well known. Our sixth-order method (with $\nu = 3$) improves on a fifth-order method of Jarratt [70].

Generalizations

Generalizations to methods using higher derivatives are possible. One result is:

Theorem 1.2

For $m > 0$, $n \geqslant 0$, and k satisfying $m + 1 \geqslant k > 0$, there exist methods which, for each iteration, use one evaluation of $f, f', \ldots, f^{(m)}$, followed by n evaluations of $f^{(k)}$, and have order of convergence $m + 2n + 1$.

The methods described here are special cases of the methods of Theorem 1.2 (take $k = m = 1$, and $\nu = n + 1)$. Since proof of Theorem 1.2 is given in Brent [75], we omit proofs here, and adopt an informal style of presentation. Other possible generalizations are mentioned in Section 7.

2. MOTIVATION

We first consider methods using one evaluation of f, and two of f', per iteration. Let x_0 be a sufficiently good approximation to the simple zero ζ of f, $f_0 = f(x_0)$, and $f'_0 = f'(x_0)$. Suppose we evaluate $f'(\tilde{x}_0)$, where

$$\tilde{x}_0 = x_0 - \alpha f_0/f'_0 ,$$

and α is a nonzero parameter. Let $Q(x)$ be the quadratic polynomial such that

$$Q(x_0) = f_0 ,$$

$$Q'(x_0) = f'_0 ,$$

and

$$Q'(\tilde{x}_0) = f'(\tilde{x}_0) ,$$

and let x_1 be the zero of $Q(x)$ closest to x_0. Jarratt [69] essentially proved:

Theorem 2.1

$$x_1 - \zeta = O(|x_0 - \zeta|^\rho)$$

as $x_0 \to \zeta$, where

$$\rho = \begin{cases} 3 & \text{if } \alpha \neq 2/3 , \\ 4 & \text{if } \alpha = 2/3 . \end{cases}$$

Thus, we choose $\alpha = 2/3$ to obtain a fourth-order method. The proof of Theorem 2.1 uses the following lemma:

Lemma 2.1

If $P(x) = a + bx + cx^2 + dx^3$ satisfies

$$P(0) = P'(0) = P'(2/3) = 0 ,$$

then $P(1) = 0$.

Applying Lemma 2.1, we may show that (for $\alpha = 2/3$)

$$f(x_N) - Q(x_N) = O(\delta^4) ,$$

where

$$x_N = x_0 - f_0/f'_0$$

is the approximation given by Newton's method, and

$$\delta = |f_0/f'_0| = |x_N - x_0| .$$

Now

$$x_N - x_1 = O(\delta^2) ,$$

and

$$f'(x) - Q'(x) = O(\delta^2)$$

for x near x_N , so

$$\begin{aligned} |f(x_1)| &= |f(x_1) - Q(x_1)| \\ &\leq |f(x_N) - Q(x_N)| + |f'(\xi) - Q'(\xi)| \cdot |x_N - x_1| \end{aligned}$$

for some ξ between x_N and x_1 . Thus

$$|f(x_1)| = O(\delta^4) + O(\delta^2 \cdot \delta^2) = O(\delta^4) ,$$

and

$$x_1 - \zeta = O(|f(x_1)|) = O(\delta^4) = O(|x_0 - \zeta|^4) .$$

3. A SIXTH-ORDER METHOD

To obtain a sixth-order method using one more derivative evaluation than the fourth-order method described above, we need distinct, nonzero parameters, α_1 and α_2 , such that

$$P(0) = P'(0) = P'(\alpha_1) = P'(\alpha_2) = 0$$

implies $P(1) = 0$, for all fifth-degree polynomials

$$P(x) = a + bx + \ldots + fx^5 .$$

Thus, we want the conditions

$$2\alpha_1 c + \ldots + 5\alpha_1^4 f = 0$$

and

$$2\alpha_2 c + \ldots + 5\alpha_2^4 f = 0$$

to imply

$$c + \ldots + f = 0 .$$

Equivalently, we want

$$\operatorname{rank}\begin{bmatrix} 2\alpha_1 & 3\alpha_1^2 & 4\alpha_1^3 & 5\alpha_1^4 \\ 2\alpha_2 & 3\alpha_2^2 & 4\alpha_2^3 & 5\alpha_2^4 \\ 1 & 1 & 1 & 1 \end{bmatrix} = 2 ,$$

i.e.,

$$\operatorname{rank}\begin{bmatrix} 1 & \alpha_1 & \alpha_1^2 & \alpha_1^3 \\ 1 & \alpha_2 & \alpha_2^2 & \alpha_2^3 \\ 1/2 & 1/3 & 1/4 & 1/5 \end{bmatrix} = 2 ,$$

i.e., for some w_1 and w_2 ,

(3.1) $$w_1\alpha_1^i + w_2\alpha_2^i = 1/(i + 2)$$

for $0 \le i \le 3$.

Since $1/(i + 2) = \int_0^1 x^i \cdot x dx$, we see from (3.1) that α_1 and α_2 should be chosen as the zeros of the Jacobi polynomial, $G_2(2, 2, x) = x^2 - 6x/5 + 3/10$, which is orthogonal to lower degree polynomials, with respect to the weight function x , on $[0, 1]$.

Let $y_i = x_0 - \alpha_i f_0/f_0'$, $x_N = x_0 - f_0/f_0'$, $\delta = |f_0/f_0'|$, and let $Q(x)$ be the cubic polynomial such that

$$Q(x_0) = f_0 , \quad Q'(x_0) = f_0' ,$$

and

$$Q'(y_i) = f'(y_i)$$

for $i=1,2$. Then

$$f(x) - Q(x) = 0(\delta^4)$$

for x between x_0 and x_N , but

$$f(x_N) - Q(x_N) = 0(\delta^6) ,$$

because of our choice of α_1 and α_2 as zeros of $G_2(2, 2, x)$.

(This might be called "superconvergence": see de Boor and Swartz [73].)

A Problem

Since

$$x_N - x_1 = O(\delta^2)$$

and

$$f'(x) - Q'(x) = O(\delta^3)$$

for x near x_N, proceeding as above gives

$$|f(x_1)| = O(\delta^6) + O(\delta^3 \cdot \delta^2) = O(\delta^5) ,$$

so the method is only of order five, not six.

A Solution

After evaluating $f'(y_1)$, we can find an approximation $\tilde{x}_N = \zeta + O(\delta^3)$ which is (in general) a better approximation to ζ than is x_N. From the above discussion, we can get a sixth-order method if we can ensure superconvergence at $\tilde{x}_N$ rather than x_N. Define $\tilde{\alpha}_1$ by

$$\tilde{\alpha}_1(\tilde{x}_N - x_0) = \alpha_1(x_N - x_0) .$$

In evaluating f' at $y_1 = x_0 + \tilde{\alpha}_1(\tilde{x}_N - x_0)$, we effectively used $\tilde{\alpha}_1 = \alpha_1 + O(\delta)$ instead of α_1, so we must perturb α_2 to compensate for the perturbation in α_1.

From (3.1), we want $\tilde{\alpha}_2$ such that, for some $\tilde{w}_1$ and $\tilde{w}_2$,

$$\tilde{w}_1\tilde{\alpha}_1^i + \tilde{w}_2\tilde{\alpha}_2^i = 1/(i + 2) \tag{3.2}$$

for $0 \le i \le 2$. Thus

$$\operatorname{rank}\begin{bmatrix} 1 & \tilde{\alpha}_1 & \tilde{\alpha}_1^2 \\ 1 & \tilde{\alpha}_2 & \tilde{\alpha}_2^2 \\ 1/2 & 1/3 & 1/4 \end{bmatrix} = 2 ,$$

which gives

$$\tilde{\alpha}_2 = (3 - 4\tilde{\alpha}_1)/(4 - 6\tilde{\alpha}_1) = \alpha_2 + 0(\delta) .$$

Since

$$\tilde{w}_j = w_j + 0(\delta)$$

for $j=1,2$, we have

$$\tilde{w}_1\tilde{\alpha}_1^3 + \tilde{w}_2\tilde{\alpha}_2^3 = 1/5 + 0(\delta) . \tag{3.3}$$

(Compare (3.1) with $i = 3$.) If we evaluate f' at $\tilde{y}_2 = x_0 + \tilde{\alpha}_2(\tilde{x}_N - x_0)$, and let x_1 be a sufficiently good approximation to the appropriate zero of the cubic which fits the data obtained from the f and f' evaluations, then (3.2) and (3.3) are sufficient to ensure that the method has order <u>six</u> after all.

4. METHODS OF ORDER 2ν

In this section we describe a class of methods satisfying Theorem 1.1. The special cases $\nu = 2$ and $\nu = 3$ have been given above.

It is convenient to define $n = \nu - 1$. The Jacobi polynomial $G_n(2, 2, x)$ is the monic polynomial, of degree n, which is orthogonal to all polynomials of degree $n - 1$, with respect to the weight function x, on $[0, 1]$. Let $\alpha_1,\ldots,\alpha_n$ denote the zeros of $G_n(2, 2, x)$ in any fixed order. We describe a class of methods of order $2(n + 1)$, using evaluations of $f(x_0)$, $f'(x_0)$, and $f'(y_1),\ldots,f'(y_n)$, where the points $y_1,\ldots,y_n$ are determined during the iteration.

<u>The Methods</u>

1. Evaluate $f_0 = f(x_0)$ and $f_0' = f'(x_0)$.
2. If $f_0 = 0$ set $x_1 = x_0$ and stop, else set $\delta = |f_0/f_0'|$.
3. For $i=1,\ldots,n$ do steps 4 to 7.

4. Let p_i be the polynomial, of minimal degree, agreeing with the data obtained so far. Let z_i be an approximate zero of p_i, satisfying $z_i = x_0 + 0(\delta)$ and $p_i(z_i) = 0(\delta^{i+2})$. (Any suitable method, e.g. Newton's method, may be used to find z_i.)

5. Compute $\alpha_{i,j} = \alpha_{i-1,j}(z_{i-1} - x_0)/(z_i - x_0)$ for $j=1,\ldots,i-1$. (Skip if $i = 1$.)

6. Let q_i be the monic polynomial, of degree $n + 1 - i$, such that $\int_0^1 P(x)\, q_i(x) \left(\prod_{j=1}^{i-1} (x - \alpha_{i,j})\right) x dx = 0$ for all polynomials P of degree $n - i$. (The existence and uniqueness of q_i may be shown constructively: see Brent [75].) Let $\alpha_{i,i}$ be an approximate zero of q_i, satisfying $\alpha_{i,i} = \alpha_i + 0(\delta)$ and $q_i(\alpha_{i,i}) = 0(\delta^{i+1})$.

7. Evaluate $f'(y_i)$, where

$$y_i = x_0 + \alpha_{i,i}(z_i - x_0) .$$

8. Let p_{n+1} be as at step 4, and x_1 an approximate zero of p_{n+1}, satisfying $x_1 = x_0 + 0(\delta)$ and $p_{n+1}(x_1) = 0(\delta^{2n+3})$.

Asymptotic Error Constants

The asymptotic error constant of a stationary zero-finding method is defined to be

$$K = \lim_{x_0 \to \zeta} (x_1 - \zeta)/(x_0 - \zeta)^\rho ,$$

where ρ is the order of convergence. (Since ρ is an integer for all methods considered here, we allow K to be signed.) Let K_ν be the asymptotic error constant of the methods (of order 2ν) described above. The general form of K_ν is not known, but we have

$$K_1 = \phi_2 ,$$

$$K_2 = \phi_4/9 - \phi_2\phi_3 ,$$

$$K_3 = \phi_6/100 + (1 - 5\alpha_1)\phi_2\phi_5/10 + (3\alpha_1 - 2)\phi_3\phi_4/5 ,$$

and

$$K_4 = \Big\{3\phi_8 - 21\phi_2\phi_7/(1 - \alpha_1) + 9[35(1 - \alpha_3) - 3/(1 - \alpha_2)]\phi_3\phi_6 - 25(9 - 44\alpha_3 + 42\alpha_3^2)\phi_4\phi_5\Big\}/3675 ,$$

where

$$\phi_i = \frac{f^{(i)}(\zeta)}{i!f'(\zeta)} .$$

5. RELATED NONLINEAR RUNGE-KUTTA METHODS

The ordinary differential equation

$$dx/dt = g(x) , \quad x(t_0) = x_0 , \tag{5.1}$$

may be solved by quadrature and zero-finding: to find $x(t_0 + h)$ we need to find a zero of

$$f(x) = \int_{x_0}^{x} \frac{du}{g(u)} - h .$$

Note that $f(x_0) = -h$ is known, and $f'(x) = 1/g(x)$ may be evaluated almost as easily as $g(x)$. Thus, the zero-finding methods of Section 4 may be used to estimate $x(t_0 + h)$, then $x(t_0 + 2h)$, etc. When written in terms of g rather than f, the methods are seen to be similar to Runge-Kutta methods.

For example, the fourth-order zero-finding methods of Section 2 (with x_1 an exact zero of the quadratic $Q(x)$) gives:

$$g_0 = g(x_0) ,$$

$$\Delta = hg_0 ,$$

$$g_1 = g(x_0 + 2\Delta/3) ,$$

and

(5.2) $$x_1 = x_0 + 2\Delta/[1 + (3g_0/g_1 - 2)^{\frac{1}{2}}] .$$

Note that (5.1) is nonlinear in g_0 and g_1 , unlike the usual Runge-Kutta methods. (This makes it difficult to generalize our methods to systems of differential equations.) Since the zero-finding method is fourth-order, $x_1 = x(t_0 + h) + O(h^4)$, so our nonlinear Runge-Kutta method has order <u>three</u> by the usual definition of order (Henrici [62]).

Similarly, any of the zero-finding methods of Section 4 have a corresponding nonlinear Runge-Kutta method. Thus, we have:

Theorem 5.1

If $\nu > 0$, there is an explicit, nonlinear, Runge-Kutta method of order $2\nu - 1$, using ν evaluations of g per iteration, for single differential equations of the form (5.1).

By the result of Meersman and Wozniakowski, mentioned in Section 1, the order $2\nu - 1$ in Theorem 5.1 is the best possible. Butcher [65] has shown that the order of *linear* Runge-Kutta methods, using ν evaluations of g per iteration, is at most ν , which is less than the order of our methods if $\nu > 1$ (though the linear methods may also be used for systems of differential equations).

6. SOME NUMERICAL RESULTS

In this section we give some numerical results obtained with the nonlinear Runge-Kutta methods of Section 5. Consider the differential equation (5.1) with

(6.1) $$g(x) = (2\pi)^{\frac{1}{2}}\exp(x^2/2)$$

and $x(0) = 0$. Using step sizes $h = 0.1$ and 0.01, we estimated $x(0.4)$, obtaining a computed value x_h . The error e_h was defined by

$$e_h = (2\pi)^{-\frac{1}{2}} \int_0^{x_h} \exp(-u^2/2)du - 0.4 .$$

All computations were performed on a Univac 1108 computer, with a floating-point fraction of 60 bits. The results are summarized in Table 6.1. The first three methods are derived from the zero-finding methods of Section 4 (with $\nu = 2, 3$ and 4 respectively). Method RK4 is the classical fourth-order Runge-Kutta method of Kutta [01], and method RK7 is a seventh-order method of Shanks [66].

Table 6.1: Comparison of Runge-Kutta Methods

Method	g evaluations per iteration	Order	$e_{0.1}$	$e_{0.01}$
Sec. 4	2	3	-9.45'-6	1.49'-7
Sec. 4	3	5	3.16'-6	-2.47'-11
Sec. 4	4	7	3.86'-8	3.69'-15
RK4	4	4	1.95'-5	7.90'-9
RK7	9	7	-5.19'-7	-1.67'-13

More extensive numerical results are given in Brent [75]. Note that the differential equation (6.1) was chosen only for illustrative purposes: there are several other ways of computing quantiles of the normal distribution. A practical application of our methods (computing quantiles of the incomplete Gamma and other distributions) is described in Brent [76].

7. OTHER ZERO-FINDING METHODS

In Section 1 we stated some generalizations of our methods (see Theorem 1.2). Further generalizations are described in Meersman [75]. Kacewicz [75] has considered methods which use information about an integral of f instead of a derivative of f.

"Sporadic" methods using derivatives may be derived as in Sections 2 and 3. For example, is there an eighth-order method which uses evaluations of f , f' , f'' , and f''' at x_0 , followed by evaluations of f' , f'' and f''' at some point y_1 ? Proceeding as in Sections 2 and 3 , we need a nonzero α satisfying

$$\text{rank} \begin{bmatrix} 1 & 1 & 1 & 1 \\ 4 & 5\alpha & 6\alpha^2 & 7\alpha^3 \\ 12 & 20\alpha & 30\alpha^2 & 42\alpha^3 \\ 24 & 60\alpha & 120\alpha^2 & 210\alpha^3 \end{bmatrix} = 3 ,$$

which reduces to

$$35\alpha^3 - 84\alpha^2 + 70\alpha - 20 = 0 . \tag{7.1}$$

Since (7.1) has one real root, $\alpha = 0.7449...$, an eighth-order method does exist. It is interesting to note that (7.1) is equivalent to the condition

$$\int_0^1 x^3(x - \alpha)^3 dx = 0 .$$

As a final example, we consider sixth-order methods using $f(x_0)$, $f'(x_0)$, $f''(y_1)$, and $f'''(y_2)$. (These could be called Abel-Gončarov methods.) Proceeding as above, we need α_1 and α_2 such that

$$\text{rank} \begin{bmatrix} 2 & 6\alpha_1 & 12\alpha_1^2 & 20\alpha_1^3 \\ 0 & 6 & 24\alpha_2 & 60\alpha_2^2 \\ 1 & 1 & 1 & 1 \end{bmatrix} = 2 ,$$

which gives

$$60\alpha_1^4 - 80\alpha_1^3 + 60\alpha_1^2 - 24\alpha_1 + 3 = 0 \tag{7.2}$$

and

$$\alpha_2 = (1 - 6\alpha_1^2)/(4 - 12\alpha_1) .$$

Fortunately, (7.2) has two real roots, $\alpha_1 = 0.2074...$ and $\alpha_1 = 0.5351...$ Choosing one of these, we may evaluate $f(x_0)$, $f'(x_0)$ and $f''(y_1)$, where y_1 is defined as in Section 3. We may then fit a quadratic to the data, compute the perturbed $\tilde{\alpha}_1$, and take

$$\tilde{\alpha}_2 = (1 - 6\tilde{\alpha}_1^2)/(4 - 12\tilde{\alpha}_1) ,$$

etc., as in Section 3. It is not known whether this method can be generalized, i.e., whether real methods of order $2n$, using evaluations of $f(x_0)$, $f'(x_0)$, $f''(y_1)$, ..., $f^{(n)}(y_{n-1})$, exist for all positive n .

8. ACKNOWLEDGEMENT

The suggestions of J.C. Butcher, R. Meersman, M.R. Osborne and J.F. Traub are gratefully acknowledged.

REFERENCES

de Boor and Swartz [73] de Boor, C.W.R. and Swartz, B., "Collocation at Gaussian points", SIAM J. Numer. Anal. 10 (1973), 582-606.

Brent [75] Brent, R.P., "Some high-order zero-finding methods using almost orthogonal polynomials", J. Austral. Math. Soc. (Ser. B) 1 (1975). Also available as "Efficient methods for finding zeros of functions whose derivatives are easy to evaluate", Dept. of Computer Science, Carnegie-Mellon Univ. (Dec. 1974).

Brent [76] Brent, R.P. and E.M., "Efficient computation of inverse distribution functions", to appear.

Brent, Winograd and Wolfe [73] Brent, R.P., Winograd, S. and Wolfe, P., "Optimal iterative processes for root-finding", Numer. Math. 20 (1973), 327-341.

Butcher [65] Butcher, J.C., "On the attainable order of Runge-Kutta methods", Math. Comp. 10 (1965), 408-417.

Henrici [62] Henrici, P., "Discrete variable methods in ordinary differential equations", Wiley, New York, 1962.

Jarratt [69] Jarratt, P., "Some efficient fourth-order multipoint methods for solving equations", BIT 9 (1969), 119-124.

Jarratt [70] Jarratt, P., "A review of methods for solving nonlinear algebraic equations in one variable", in "Numerical methods for nonlinear algebraic equations", (edited by P. Rabinowitz), Gordon and Breach, New York, 1970, 1-26.

Kacewicz [75] Kacewicz, B., "An integral-interpolatory iterative method for the solution of nonlinear scalar equations", Dept. of Computer Science, Carnegie-Mellon Univ. (Jan. 1975).

Kung and Traub [73] Kung, H.T. and Traub, J.F., "Optimal order and efficiency for iterations with two evaluations", Dept. of Computer Science, Carnegie-Mellon Univ. (Nov. 1973). To appear in SIAM J. Num. Anal.

Kung and Traub [74] Kung, H.T. and Traub, J.F., "Optimal order of one-point and multipoint iteration", J. ACM 21 (1974), 643-651.

Kutta [01] Kutta, W., "Beitrag zur näherungsweisen Integration totaler Differentialgleichungen", Z. Math. Phys. 46 (1901), 435-452.

Meersman [75] Meersman, R., "Optimal use of information in certain iterative processes", these Proceedings.

Shanks [66] Shanks, E.B., "Solutions of differential equations by evaluations of functions", Math. Comp. 20 (1966), 21-38.

Traub [64] Traub, J.F., "Iterative methods for the solution of equations", Prentice-Hall, Englewood Cliffs, New Jersey (1964).

Wozniakowski [75a] Wozniakowski, H., "Properties of maximal order methods for the solution of nonlinear equations", ZAMM 55 (1975), 268-271.

Wozniakowski [75b] Wozniakowski, H., "Maximal order of multipoint iterations using n evaluations," these Proceedings.

MAXIMAL ORDER OF MULTIPOINT ITERATIONS USING n EVALUATIONS*

H. Woźniakowski
Department of Computer Science
Carnegie-Mellon University
(On leave from University of Warsaw)

ABSTRACT

This paper deals with multipoint iterations without memory for the solution of the nonlinear scalar equation $f^{(m)}(x) = 0$, $m \geq 0$. Let $p_n(m)$ be the maximal order of iterations which use n evaluations of the function or its derivatives per step. We prove the Kung and Traub conjecture $p_n(0) = 2^{n-1}$ for Hermitian information. We show $p_n(m+1) \geq p_n(m)$ and conjecture $p_n(m) \equiv 2^{n-1}$. The problem of the maximal order is connected with Birkhoff interpolation. Under a certain assumption we prove that the Pólya conditions are necessary for maximal order.

1. INTRODUCTION

We consider the problem of solving the nonlinear scalar equation $f^{(m)}(x) = 0$ where m is a nonnegative integer. We solve this problem by multipoint iterations without memory which use n evaluations of the function or its derivatives per step. For fixed n we seek an iteration of maximal order of convergence. This problem is connected with Birkhoff interpolation and can be expressed in terms of the incidence matrix $E_n^k = (e_{ij})$ where $e_{ij} = 1$ if $f^{(j)}(z_i)$ is computed and

$e_{ij} = 0$ otherwise; $z_i \neq z_j$, and $\sum_{i=1}^{k} \sum_{j=0}^{\infty} e_{ij} = n$. (Note that the problem of Birkhoff interpolation has been open for 70 years, see Sharma [72].)

Let $p_n(m)$ be the maximal order of multipoint iterations. For $m = 0$, Kung and Traub showed that $p_n(0) \geq 2^{n-1}$. We show that $p_n(m+1) \geq p_n(m)$ and conjecture $p_n(m) = 2^{n-1}$. For $m = 0$ we prove the Kung and Traub conjecture for Hermitian information, i.e., if $f^{(j)}(z_i)$ is computed, then $f^{(0)}(z_i),\dots,f^{(j-1)}(z_i)$ are also computed. Under a certain assumption we prove that the Pólya conditions are necessary for the maximal order, i.e., the total number of $f, f', \dots, f^{(j)}$ evaluations has to be at least $j+1$, $j = 0,1,\dots,n-1$. We show also that $p_n(0) \leq n(n+1)^{n-1}$. Some special incidence matrices E_n^k are considered and maximal orders of iterations based on E_n^k are discussed.

2. THE n-EVALUATION PROBLEM

We consider the problem of solving the nonlinear scalar equation

$$(2.1) \quad f^{(m)}(x) = 0$$

where $f: D_F \subset \mathbb{C} \to \mathbb{C}$, $\mathbb{C}$ denotes the one dimensional complex space and m is a nonnegative integer. We assume that there exists a simple zero α of $f^{(m)}$, $f^{(m)}(\alpha) = 0 \neq f^{(m+1)}(\alpha)$, and that f is analytic in a neighborhood of α. Let $\mathfrak{F}$ denote a class of such functions.

We solve (2.1) by stationary iteration and assume that x_1 is a sufficiently close approximation to α. To get the next approximation x_2 to α we need some information on f. We assume that this information $\mathfrak{N} = \mathfrak{N}(x_1; f)$ is given by some values of the function and its derivatives at the points z_i defined as follows. Let

$$\begin{aligned} z_1 &: f^{(j_1^1)}(z_1),\dots,f^{(j_{\mu_1}^1)}(z_1), \\ &\vdots \\ z_k &: f^{(j_1^k)}(z_k),\dots,f^{(j_{\mu_k}^k)}(z_k) \end{aligned}$$

denote points and numbers of derivatives which are computed where nonnegative integers $\{j_\mu^i\}$ satisfy the relations

$$j_\mu^i < j_{\mu+1}^i \quad \text{for } i=1,2,\dots,k \text{ and } \mu=1,2,\dots,\mu_i-1,$$

$$\mu_1 + \mu_2 + \dots + \mu_k = n.$$

Furthermore,

$$\begin{aligned} &z_1 = x_1 \\ (2.2)\quad &z_{i+1} = z_{i+1}(z_1,\dots,z_i,\ f^{(j_1^1)}(z_1),\dots,\ f^{(j_{\mu_1}^1)}(z_1),\dots, \\ &\qquad f^{(j_1^i)}(z_i),\dots,f^{(j_{\mu_i}^i)}(z_i)) \quad \text{for } i = 1,2,\dots,k, \end{aligned}$$

$$z_i \neq z_j \text{ for } x_1 \neq \alpha \text{ and } i \neq j,\ i,j = 1,2,\dots,k,$$

$$x_2 = z_{k+1}.$$

This means that every z_{i+1} is the function of the previous information computed at $z_1,\dots,z_i$ and the next approximation $x_2 = z_{k+1}$ depends on n evaluations. Sometimes we shall use the notation $z_i = z_i(x_1)$ or $z_i = z_i(x_1,f)$ to stress the dependence on x_1 and f.

To simplify further notations we define _an incidence matrix_ $E_n^k = (e_{ij})$ _of the information_ $\mathfrak{N}$, $i = 1,2,\dots,k$ and $j = 0,1,\dots,$ as follows. Let

$$(2.3)\quad e_{ij} = \begin{cases} 1 & \text{if we compute } f^{(j)}(z_i) \\ 0 & \text{if we do not compute } f^{(j)}(z_i), \end{cases}$$

where

$$(2.4)\quad \sum_{j=0}^{\infty} e_{ij} > 0 \qquad \text{for } i = 2,3,\dots,k,$$

$$(2.5)\quad |E_n^k| = \sum_{i=1}^{k} \sum_{j=0}^{\infty} e_{ij} = n, \text{ (thus } k \le n+1).$$

The condition (2.4) means that at every point z_i, $i \ge 2$, we compute at least one derivative. (We consider f to be the zeroth derivative $f^{(0)}$.) However we do not, at this point, insist on any information being computed at $z_1 = x_1$. We show in Lemma 3.2 that $f^{(m)}$ must be evaluated at x_1. The condition (2.5) means that we use exactly n evaluations. Let

$$(2.6)\quad e_n^k = \{(i,j): e_{ij} = 1,\ i = 1,2,\dots,k;\ j = 0,1,\dots\}$$

Hence the information $\mathfrak{N}$ can then be defined in terms of the incidence matrix E_n^k as follows:

$$(2.7)\quad \mathfrak{N} = \mathfrak{N}(x_1;f) = \{f^{(j)}(z_i) : (i,j) \in e_n^k\}.$$

The concept of an incidence matrix is used in Birkhoff interpolation, see Sharma [72]. We shall show some connections between the n evaluation problem and Birkhoff interpolation.

Having the information $\mathfrak{N}$ we define the next approximation x_2, $x_2 = z_{k+1}$, as $x_2 = \varphi(x_1; \mathfrak{N}(x_1;f))$ where φ is a given function.

We call φ <u>an iteration function</u> if for every $f \in \mathfrak{F}$, with $f^{(m)}(\alpha) = 0$ there exists $\delta > 0$ such that for any x_1, $|x_1-\alpha| \le \delta$, the sequence

(2.8a) $x_{d+1} = \varphi(x_d;\ \mathfrak{N}(x_d;f)),\quad d = 1,2,\ldots$

is well-defined and

(2.8b) $\lim_{d\to\infty} x_d = \alpha,$

(2.8c) $\alpha = \varphi(\alpha,\ \mathfrak{N}(\alpha;f)).$

Such iterations are called k-point iteration without memory since they use exactly n new evaluations at k distinct points. If $k > 1$ they are called <u>multipoint iterations</u> (see Traub [61], [64], and Kung and Traub [74]). Let Φ be a class of iterations φ with $k \geq 1$.

Since these iterations are stationary and without memory it is sufficient to define how x_2 is generated from x_1 and to measure the goodness of φ by examining some properties of $x_2 - \alpha$ as x_1 tends to α.

We want to find an iteration for which x_2 approximates α as closely as possible, i.e., we seek an iteration with the maximal order. In a previous paper (Wozniakowski [75]) we proved that if a set of iterations Φ is not empty then the maximal order of iteration is equal to the order of information. This gives us a powerful technique for proving maximal order. Let us briefly recall what we mean by orders of iteration and information.

We shall say <u>$\{\tilde{f}(\cdot;\ x_1)\}$ is equal to f with respect to $\mathfrak{N}$</u> (briefly denoted by $\tilde{f} \underset{\mathfrak{N}}{\equiv} f$) iff

(i) $f,\ \tilde{f}(\cdot;\ x_1) \in \mathfrak{J},$

(ii) $\tilde{f}^{(m)}(\tilde{\alpha};\ x_1) = 0$ and $f^{(m)}(\alpha) = 0$ where $\tilde{\alpha} = \tilde{\alpha}(x_1)$ and $\lim_{x_1\to\alpha} \tilde{\alpha}(x_1) = \alpha,$

(iii) $\lim\limits_{x_1 \to \alpha} \tilde{f}^{(j)}(\alpha; x_1) = g^{(j)}(\alpha)$ where $g(\alpha) = 0$ and $g \in \mathfrak{I}$, $j = 0,1,\ldots$

(iv) $\mathfrak{N}(x_1; \tilde{f}) = \mathfrak{N}(x_1; f)$, i.e., $\tilde{f}^{(j)}(z_i; x_1) = f^{(j)}(z_i)$ for $(i,j) \in e_n^k$.

The first three conditions mean that $\tilde{f}(x; x_1)$ is sufficiently regular with respect to x and tends to a function g, $g \in \mathfrak{I}$, as x_1 tends to α. The condition (iv) means that $\tilde{f}$ and f have the same information $\mathfrak{N}$ at the point x_1. Therefore any iteration φ will produce the same approximation x_2 for $\tilde{f}$ and f, $\varphi(x_1; \mathfrak{N}(x_1; \tilde{f})) \equiv \varphi(x_1; \mathfrak{N}(x_1; f))$. Since we cannot recognize $\tilde{f}$ from f using information (2.7), we should approximate not only the zero α of f, but at the same time, the zero $\tilde{\alpha}$ of $\tilde{f}$. This leads us to the following definitions of orders of iteration and information.

Let A be a set defined by

$$A = \{q \geq 1;\ \forall f \in \mathfrak{I}, f^{(m)}(\alpha) = 0,\ \forall \tilde{f} \underset{\mathfrak{N}}{\equiv} f,\ \limsup_{x_1 \to \alpha} \frac{|x_2 - \tilde{\alpha}|}{|x_1 - \alpha|^{q-\epsilon}} = 0,\ \forall \epsilon > 0\}$$

A number $p = p(\varphi)$ is called <u>an order of the iteration φ</u> iff

$$(2.9)\quad p(\varphi) = \begin{cases} 0 & \text{if A is empty,} \\ \sup A & \text{otherwise.} \end{cases}$$

Using this convention $p(\varphi)$ always exists; however the only interesting cases are for $A \neq \emptyset$. Furthermore, let

$$B = \{q \geq 1;\ \forall f \in \mathfrak{I}, f^{(m)}(\alpha) = 0, \forall \tilde{f} \underset{\mathfrak{N}}{\equiv} f, \limsup_{x_1 \to \alpha} \frac{|\alpha - \tilde{\alpha}|}{|x_1 - \alpha|^{q-\epsilon}} = 0, \forall \epsilon > 0\}.$$

A number $p = p(\mathfrak{N})$ (sometimes denoted $p = p(E_n^k)$) is called <u>an order of the information $\mathfrak{N}$</u> if

$$(2.10)\quad p(\mathfrak{N}) = \begin{cases} 0 & \text{if B is empty,} \\ \sup B & \text{otherwise.} \end{cases}$$

We know that if $\Phi \neq \emptyset$ then

$$(2.11)\quad \sup_{\varphi \in \Phi} p(\varphi) = p(\mathfrak{N})$$

and $p(\mathfrak{N}) = p(I_{\mathfrak{N}})$ where $I_{\mathfrak{N}}$ is a generalized interpolatory method. (See Wozniakowski [75].)

We are now in a position to define the n-evaluation problem (see Kung and Traub [73] and [74]). For fixed n and m we wish to find a number $k = k(n,m)$, points $z_i = z_i(x_1)$ for $i = 2,3,\ldots,k$, an incidence matrix E_n^k, $|E_n^k| = n$, and an iteration φ which uses E_n^k (see (2.8)) such that $p(\varphi)$ is maximal. Due to (2.11) this is equivalent to maximizing the order of information $\mathfrak{N}$, i.e., to find $E*_n^k$ such that

$$(2.12)\quad p_n(m) = \sup_{E_n^k} p(E_n^k),$$

$$(2.13)\quad p(E*_n^k) = p_n(m).$$

We recall the Kung and Traub conjecture for m = 0 (Kung and Traub [74]):

$$(2.14)\quad p_n(0) = 2^{n-1}.$$

They showed two different matrices E_n^k, $n \geq 2$, for which the order of iteration is equal to 2^{n-1} (see Section 3), so we know that

$$(2.15)\quad p_n(0) \geq 2^{n-1}.$$

We now show a relationship among the $p_n(m)$ for different m.

Lemma 2.1

Let $\varphi = \varphi(\mathfrak{N})$ be an iteration of order p for the problem $f^{(m)}(x) = 0$ which uses n evaluations per step. Then there exists an iteration $\varphi^* = \varphi^*(\mathfrak{N}^*)$ for the problem $f^{(m+1)}(x) = 0$ which also uses n evaluations and has the same order p.

Proof

Let $E_n^k = (e_{ij})$ be the incidence matrix of $\mathfrak{N}$ and $E_n^{*k} = (e_{ij}^*)$ be defined by

$$e_{ij}^* = \begin{cases} 1 & \text{if } e_{i,j-1} = 1 \\ 0 & \text{otherwise.} \end{cases}$$

Let $\mathfrak{N}^*$ be information with the incidence matrix E_n^{*k} based on the points $z_i = z_i(x_1)$, $i = 2,\ldots,k$, from $\mathfrak{N}$. For any f_1 from $\mathfrak{F}$, $f_1^{(m+1)}(\alpha) = 0 \neq f_1^{(m+2)}(\alpha)$, define

$$f(x) = f_1'(x).$$

Thus, $f \in \mathfrak{F}$, $f^{(m)}(\alpha) = 0 \neq f^{(m+1)}(\alpha)$, and $f^{(j)}(x) \equiv f_1^{(j+1)}(x)$. Hence

$$\mathfrak{N}^*(x_1;\ f_1) = \mathfrak{N}(x_1;\ f).$$

Let us define φ^* by

$$\varphi^*(x_1;\ \mathfrak{N}^*(x_1;\ f_1)) = \varphi(x_1,\ \mathfrak{N}(x_1;\ f)).$$

Since f_1 is arbitrary it easily follows that $p(\varphi^*) = p(\varphi)$. ∎

From Lemma 2.1 and (2.15) we immediately get

Corollary 2.2

$p_n(m) \geq p_n(m-1) \geq 2^{n-1}$ for any $m \geq 1$. ∎

Although Corollary 2.2 states that $p_n(m)$ is at least $p_n(m-1)$ we propose

Conjecture 2.3

$$p_n(m) = 2^{n-1} \qquad \forall\, m \geq 0,\ n \geq 1. \qquad \blacksquare$$

3. EXISTENCE OF ITERATIONS

Recall that Φ is a class of iterations defined by (2.8). In this section we show what we have to assume on the information $\mathfrak{N}$ to be sure that Φ is not empty. We shall prove that $\Phi = \emptyset$ if any of the following three conditions hold:

(1) If $z_i(x_1)$ does not converge to α.

(2) If we do not compute $f^{(m)}(x_1)$, i.e., $e_{1m} = 0$.

(3) If $n = 1$ under the assumption on sufficiently regularity of φ as a function of x_1.

We prove this in the following Lemmas.

Lemma 3.1

Let φ be an iteration which uses the information $\mathfrak{N}$. Then for any $f \in \mathfrak{F}$, $f^{(m)}(\alpha) = 0$,

$$\lim_{x_1 \to \alpha} z_i(x_1;\ f) = \alpha \qquad \text{for } i = 1,2,\ldots,k+1.$$

Proof

Suppose on the contrary that there exist $f \in \mathfrak{F}$, $f(\alpha) = 0$, an index i, $2 \leq i \leq k$, a number $\epsilon > 0$ and a sequence $\{x_j\}$ such that

$$\lim_{j\to\infty} x_j = \alpha \quad \text{and} \quad |z_i(x_j) - \alpha| \geq \epsilon \quad \text{for } j \geq j_0.$$

Let $J = \{x\colon |x-\alpha| < \epsilon\}$. Define $f_1\colon J \to \mathbb{C}$ such that $f_1(x) = f(x)$ for $x \in J$. Since $f_1 \in \mathfrak{F}$ there exists $\delta_1 > 0$ such that any x_1, $|x_1 - \alpha| \leq \delta_1$ is a good initial approximation.

Setting $x_1 = x_j$, for large j, where $|x_j-\alpha| \le \delta_1$, we get $z_i(x_j) \notin J$ and $\mathfrak{N}(x_1;\, f_1)$ is not well defined which contradicts (2.8a). ■

Lemma 3.2

Let $\mathfrak{N}$ be any information with the incidence matrix E_n^k. If $\Phi \neq \emptyset$ then $e_{1m} = 1$, (i.e. we have to compute $f^{(m)}(x_1)$).

Compare Theorem 4.1 in Kung and Traub [73] which proves this result for m = 0.

Proof

Let $\varphi \in \Phi$ and suppose on the contrary that $e_{1m} = 0$. Let f be any function from $\mathfrak{I}$, $f^{(m)}(\alpha) = 0$. Let x_1 be sufficiently close approximations to α, $x_1 \neq \alpha$. From (2.2) we get $\delta = \min_{2 \le i \le k} |z_i(x_1) - x_1| > 0$.

Define

$$f_1(x) = \begin{cases} f(x) - \dfrac{f^{(m)}(x_1)}{m!}(x-x_1)^m & \text{for } |x-x_1| < \delta \\ f(x) & \text{otherwise} \end{cases}$$

Note that $f_1 \in \mathfrak{I}$, $f_1^{(m)}(x_1) = 0$, and

$$f_1^{(j)}(x_1) = f^{(j)}(x_1) \qquad \text{for } j \neq m$$

$$f_1^{(j)}(z_i) = f^{(j)}(z_i) \qquad \text{for any j and } i = 2,\ldots,k.$$

Since we do not compute $f^{(m)}(x_1)$ then

$$\mathfrak{N}(x_1;\, f_1) = \mathfrak{N}(x_1;\, f).$$

But x_1 is the zero of f_1 and due to (2.8c) it follows

$$x_2 = \varphi(x_1;\, \mathfrak{N}(x_1;f)) = \varphi(x_1;\, \mathfrak{N}(x_1;f_1)) = x_1.$$

Thus, $x_d \equiv x_1$ and $\lim_d x_d \neq \alpha$ which contradicts (2.8b). ■

An iteration function φ can be treated as a function of x, $\varphi(x) = \varphi(x; \mathfrak{N}(x; f))$ for x close to α. We shall prove that if φ is sufficiently regular then the number of evaluations n has to be at least two.

Lemma 3.3

If an iteration φ is a sufficiently smooth function of x then $n \geq 2$.

Proof

It is enough to prove Lemma 3.3 for the real case. Assume on the contrary that $n = 1$. From Lemma 3.2 it follows that this unique piece of information is given by $f^{(m)}(x_1)$. Let

$$\varphi(x; f^{(m)}(x)) = x + g(x, f^{(m)}(x)).$$

From (2.8b) it follows

$$g(\alpha, 0) = 0 \qquad \forall\, \alpha \text{ such that } f^{(m)}(\alpha) = 0,\ f \in \mathfrak{J}$$

From this and the regularity of φ we can express $g(x, y)$

$$g(x, y) = y^k h(x, y)$$

for an integer $k \geq 1$ where $h(x, 0) \not\equiv 0$ and $h(x) = h(x, f(x))$ is a continuous function for x close to α.

Let $h(\alpha) \neq 0$ and for simplicity we assume that $h(\alpha) > 0$. (If $h(\alpha) < 0$ then the proof is analogous.) Let $f \in \mathfrak{J}$ be a polynomial of degree $m+1$ and $f^{(m+1)}(x) \equiv 1$, $f(\alpha) = 0$. There exists $\delta = \delta(f) > 0$ such that for any x_1, $|x_1 - \alpha| \leq \delta$ the sequence $x_{d+1} = \varphi(x_d, f^{(m)}(x_d)) = x_d + \left[f^{(m)}(x_d)\right]^k h(x_d)$ is well defined for any d and converges to α (see (2.8)). For

$e_d = x_d - \alpha$ we get

(3.2) $e_{d+1} = [1 + e_d^{k-1}\, h(x_d)]\, e_d.$

If x_1 is close but different from α then $e_d \neq 0$ for any d. Since $\lim e_d = 0$ then for any d_1 there exists $d \geq d_1$ such that $|e_{d+1}|^d < |e_d|$, i.e.

(3.3) $|1 + e_d^{k-1}\, h(x_d)| < 1.$

We consider two cases.

Case I. Let k be odd. Then for large d we have

$$e_d^{k-1}\, h(x_d) \cong e_d^{k-1}\, h(\alpha) > 0$$

which contradicts (3.3).

Case II. Let k be even. We prove that h does not change sign for $x \in [\alpha-\delta, \alpha+\delta]$. If so, then by the continuity of h there exists x^* such that $h(x^*) = 0$ and $0 < |x^*-\alpha| < \delta$. Setting $x_1 = x^*$ we get $x_d \equiv x^*$ which contradicts (3.3). Thus $h(x) \geq h_0 > 0$ for $|x-\alpha| \leq \delta$. Define $f_1\colon [\alpha-\delta, \alpha+\delta] \to \mathfrak{R}$ such that $f_1(x) = f(x)$. Since f_1 also belongs to $\mathfrak{F}$, $f_1^{(m)}(\alpha) = 0$, there exists $\delta_1 > 0$ such that $x_{d+1} = \varphi(x_d;\ \mathfrak{N}(x_d;\ f_1))$ is well defined whenever $|x_1-\alpha| \leq \delta_1$. Let $x_1 > \alpha$. Keeping in mind that $\mathfrak{N}(x_d;\ f_1) \equiv \mathfrak{N}(x_d;\ f)$, from (3.2) we get

$$e_{d+1} \geq (1 + e_d^{k-1} h_0) e_d \geq (1 + e_1^{k-1} h_0)^d e_1.$$

Hence, there exists an index d such that $e_{d+1} > \delta$, and since $f_1(x_{d+1})$ is not defined we get a contradiction with (2.8a). ∎

4. HERMITIAN INFORMATION

In this section we deal with a special case of the n-evaluation problem when the information $\mathfrak{N}$ is hermitian.

Definition 4.1

$\mathfrak{N}$ is called hermitian information if the incidence matrix E_n^k (which is now called hermitian) satisfies

$$e_{ij} = 1 \Rightarrow e_{i0} = e_{i1} = \dots = e_{i,j-1} = 1 \quad \forall\ (i,j) \in e_n^k \quad \blacksquare$$

This means that if $f^{(j)}(z_i)$ is computed then $f^{(0)}(z_i),\dots,$ $f^{(j-1)}(z_i)$ are also computed.

Let s_i denote the number of evaluations at z_i, i.e., $e_{i,s_i-1} = 1$ and $e_{i,s_i} = 0$. Then

$$(4.1) \quad s_1 + s_2 + \dots + s_k = n \text{ where } s_i \geq 1 \text{ for } i = 1,2,\dots,k.$$

For given n and k we want to find s_i and z_i, $i = 1,2,\dots,k$, to maximize the order of information. Let $p_n(m,H)$ be the maximal order of hermitian information. Note that $p_n(m) \geq p_n(m,H)$.

First we shall discuss a property of hermitian informations for the problem $f(x) = 0$, i.e., $m = 0$.

Theorem 4.1 $(m = 0)$

The order $p(E_n^k)$ of the hermitian information $\mathfrak{N}$ with the incidence matrix E_n^k satisfies

$$(4.2) \quad p(E_n^k) \leq s_1 \prod_{i=2}^{k} (s_i+1). \quad \blacksquare$$

Proof

It is easy to verify that if $\tilde{f} \underset{\mathfrak{N}}{=} f$ then

$$(4.3)\quad \tilde{f}(x;\, x_1) = f(x) + G(x;\, x_1) \prod_{i=1}^{k} (x-z_i)^{s_i}$$

for an analytic function G. Since $\tilde{f}'(\alpha;\, x_1)$ tends to $g'(\alpha) \neq 0$ then setting $x = \alpha$ in (4.3) we get

$$(4.4)\quad (\alpha-\tilde{\alpha}) = \frac{G(\alpha;\, x_1)}{g'(\alpha)}(1 + o(1)) \prod_{i=1}^{k} (\alpha-z_i)^{s_i}.$$

Define q_i by

$$\frac{\alpha-z_i}{e_1^{q_i-\varepsilon}} \to 0 \quad \text{and} \quad \frac{\alpha-z_i}{e_1^{q_i+\varepsilon}} \to +\infty, \quad \forall\varepsilon > 0$$

where $e_1 \equiv x_1 - \alpha$. Since $z_i = z_i(x_1)$ tends to α (see Lemma 3.1) then q_i exists and $q_i \geq 0$ for $i = 1,2,\ldots,k$. Note that $q_1 = 1$.

Let $p_1 = q_1 = 1$ and

$$(4.5)\quad p_{j+1} = \sum_{i=1}^{j} q_i s_i, \quad j = 1,2,\ldots,k.$$

From (4.4) we get

$$(4.6)\quad \frac{\alpha-\tilde{\alpha}}{e_1^{p_{k+1}-\varepsilon}} = \frac{G(\alpha;\, x_1)}{g'(\alpha)}(1 + o(1)) \prod_{i=1}^{k} \left\{\frac{\alpha-z_i}{e_1^{q_i-\delta}}\right\}^{s_i} \to 0, \ \forall\varepsilon>0,$$

where $\delta = \varepsilon/n$. For $G(\alpha;\, x_1) \equiv \text{const} \neq 0$ we get

$$(4.7)\quad \frac{\alpha-\tilde{\alpha}}{e_1^{p_{k+1}+\varepsilon}} \to \infty, \ \forall\varepsilon > 0.$$

Now we shall prove that there exists a function f such that

(4.8) $q_i \le p_i$ for $i = 1,2,\dots,k$.

Let f be any function such that $f \in \mathfrak{J}$, $f(\alpha) = 0$ and $f^{(j)}(\alpha) \neq 0$ for $j = 1,2,\dots$. Since $p_1 = q_1$, the condition (4.8) holds for $i = 1$. Assume by induction that this holds for $i \le j$. Suppose by the contrary that

$$q_{j+1} > p_{j+1} = \sum_{i=1}^{j} q_i s_i.$$

Define

$$r = \sum_{i=1}^{j} s_i.$$

Case I. Let $r = 1$. This means that $j = 1$, $s_1 = 1$ and $z_2 = z_2(x_1, f(x_1))$ approximates α with order greater than $p_2 = 1$.

Define

$$h(x_1, f(x_1)) = \frac{x_1 - f(x_1) - z_2}{z_2 - x_1} + 1. \tag{4.9}$$

It is easy to verify that

$$h(x_1, f(x_1)) = f'(\alpha)(1 + o(1)).$$

Case II. Let $r > 1$ and $\tilde{f}$ be the Hermite interpolatory polynomial of degree less than r defined by

$$\tilde{f}^{(l)}(z_i) = f^{(l)}(z_i), \quad i = 1,2,\dots,j;\ l = 0,1,\dots,s_i-1.$$

Let $\tilde{\alpha}$ be the nearest zero of $\tilde{f}$ to $z_1 = x_1$. Then

$$(4.10)\quad \frac{\tilde{\alpha} - \alpha}{\prod_{i=1}^{j} (\alpha - z_i)^{s_i}} f'(\alpha) = \frac{f^{(r)}(\alpha)}{r!}(1 + o(1)).$$

Note that $\tilde{\alpha}$ is a function of x_1 and information $\mathfrak{N}(x_1; f) = \{f^{(l)}(z_i): i = 1,2,\ldots,j;\ l = 0,1,\ldots,s_i-1\}$. Recall that $z_{j+1} = z_{j+1}(x_1, \mathfrak{N}(x_1; f))$ and $z_{j+1} - \alpha = o(e_1^{p_{j+1}})$. Define

$$(4.11)\quad h(x_1, \mathfrak{N}(x_1; f)) = \frac{\tilde{\alpha} - z_{j+1}}{\prod_{i=1}^{j} (z_{j+1} - z_i)^{s_i}} \tilde{f}'(z_{j+1}).$$

Thus h is the lefthand side of (4.10) where α is replaced by z_{j+1}. Since z_{j+1} is a better approximation to α than $\tilde{\alpha}$, it is straightforward to verify that

$$(4.12)\quad h(x_1, \mathfrak{N}(x_1; f)) = \frac{f^{(r)}(\alpha)}{r!}(1 + o(1)).$$

This means that in both cases using r evaluations of the function and its derivatives given by $\mathfrak{N}$ we can approximate the rth normalized derivative. We prove that this is impossible.

Note that h (see (4.9) or (4.11)) is a continuous function of x_1 at $x_1 = \alpha$ and

$$(4.13)\quad h(\alpha, \mathfrak{N}(\alpha; f)) = \frac{f^{(r)}(\alpha)}{r!}.$$

Let $f_1(x) = f(x) + (x-\alpha)^r$ and let us apply h to the function f_1. Thus

$$h(\alpha, \mathfrak{N}(\alpha; f)) = h(\alpha, \mathfrak{N}(\alpha; f_1)) = \frac{f^{(r)}(\alpha)}{r!} + 1$$

which contradicts (4.13).

Hence $q_{j+1} \le p_{j+1}$ which proves (4.8). Keeping in mind $p(E_n^k) = p_{k+1}$ and using (4.5), (4.8) we get

$$p(E_n^k) = \sum_{i=1}^{k} q_i s_i \le \sum_{i=1}^{k} p_i s_i = \sum_{i=1}^{k-1} p_i s_i + p_k s_k \le (1+s_k) \sum_{i=1}^{k-1} p_i s_i$$

$$\le s_1 \prod_{i=2}^{k} (s_i+1)$$

which proves Theorem 4.1. ■

We want to show that a bound in (4.2) is sharp, i.e., there exist points $z_2,\ldots,z_k$ such that the order of information is equal to $s_1 \prod_{i=1}^{k} (s_i+1)$.

Let w_μ, $\mu = 1,2,\ldots,k$, be the Hermite interpolatory polynomial of degree less than $r_\mu = s_1 + s_2 + \ldots + s_\mu$ defined by

$$(4.14) \quad w_\mu^{(j)}(z_i) = f^{(j)}(z_i), \; i = 1,2,\ldots,\mu; \; j = 0,1,\ldots,s_i-1.$$

Let α_μ be the nearest zero of w_μ to $z_1 = x_1$. (If $s_1 = 1$ then $\alpha_1 = x_1 - \beta f(x_1)$ for any nonzero constant β.)

Define $z_{\mu+1}$ as a point such that

$$(4.15) \quad z_{\mu+1} = \alpha_\mu + O(e_1^{\beta_\mu}), \; \beta_\mu \ge s_1 \prod_{i=2}^{\mu} (s_i+1).$$

From (4.14) it follows

$$(4.16) \quad \alpha_\mu - \alpha = \begin{cases} (\beta f'(\alpha)-1)(\alpha-z_1) + o(\alpha-z_1) & r_\mu = 1 \\ \dfrac{f^{(r_\mu)}(\alpha)}{r_\mu!f'(\alpha)} \prod_{i=1}^{\mu} (\alpha-z_i)^{s_i} + o(\prod_{i=1}^{\mu} (\alpha-z_i)^{s_i}) & \text{if } r_\mu > 1. \end{cases}$$

From (4.15) we get

$$(4.17)\quad z_{\mu+1} - \alpha = O(e_1^{q_{\mu+1}}), \quad q_{\mu+1} = s_1 \prod_{i=2}^{\mu} (s_i+1),$$

which proves that the order of information $\mathfrak{N}$ based on the points $z_{\mu+1}$ from (4.15) is equal to $s_1 \prod_{i=2}^{k} (s_i+1)$.

An iteration which uses this information $\mathfrak{N}$ and has the maximal order can be defined as follows.

For $\mu = 1,2,\ldots,k$

(i) construct w_μ from (4.14) using a divided-difference algorithm,

(ii) apply Newton iteration to the equation $w_\mu(x) = 0$ setting

$$y_0 = z_\mu$$
$$y_{i+1} = y_i - w'_\mu(y_i)^{-1} w_\mu(y_i), \quad i = 0,1,\ldots,i_0-1,$$
$$z_{\mu+1} = y_{i_0}$$

where

$$(4.18)\quad i_0 = \lceil \log_2(s_{\mu+1}+1) \rceil.$$

(If $s_1 = 1$ then $z_2 = x_1 - \beta f(x_1)$.)

Then (4.15) holds and

$$(4.19)\quad z_{k+1} - \alpha = O(e_1^{q_{k+1}}), \quad q_{k+1} = s_1 \prod_{i=2}^{k} (s_i+1).$$

Furthermore if $\beta_\mu > q_{\mu+1}$ in (4.15) then we can specify the constant which appears in the "O" notation in (4.19). Note that $\beta_\mu > q_{\mu+1}$ if we redefine i_0 in (4.18) as the smallest integer such that $i_0 > \log_2(s_{\mu+1}+1)$.

<u>Lemma 4.2</u>

Let φ be the iteration defined as above, $z_{k+1} = \varphi(x_1, \mathfrak{N}(x_1; f))$. If $\beta_\mu > q_{\mu+1}$ for $\mu = 1,2,\ldots,k$ then

$$(4.20)\quad \lim_{x_1\to\alpha} \frac{z_{k+1}(x_1) - \alpha}{(x_1-\alpha)^{q_{k+1}}} = C_{k+1}$$

where

$$C_{\mu+1} = M_{r_\mu} \prod_{j=1}^{\mu-1} M_{r_j}^{\,s_{j+1}(s_{j+2}+1)\ldots(s_\mu+1)} \quad \text{for } \mu = 1,2,\ldots,k$$

and

$$M_i = \begin{cases} (-1)^i \dfrac{f^{(i)}(\alpha)}{i!f'(\alpha)} & \text{if } i > 1 \\ -\beta f'(\alpha) + 1 & \text{if } i = 1. \end{cases}$$

If

$$(4.21)\quad \underline{K}^{i-1} \le \left|\frac{f^{(i)}(\alpha)}{i!f'(\alpha)}\right| \le \bar{K}^{i-1} \quad \text{for } i = r_1, r_2, \ldots, r_k$$

then

$$(4.22)\quad c\cdot \underline{K}^{q_{k+1}-1} \le \lim_{x_1\to\alpha} \left|\frac{z_{k+1}(x_1)-\alpha}{(x_1-\alpha)^{q_{k+1}}}\right| \le \bar{K}^{q_{k+1}-1}\cdot c$$

where

$$c = \begin{cases} 1 & \text{if } r_1 > 1 \\ |M_1|^{s_2(s_3+1)\ldots(s_k+1)} & \text{if } r_1 = 1 \text{ and } k \ge 2 \\ |M_1| & \text{if } r_1 = 1 \text{ and } k = 1 \end{cases}$$ ■

Note that the righthand side of (4.21) follows from the analyticity of f.

Proof

Let $C_i = \lim_{x_1 \to \alpha} (z_i-\alpha)/(x_1-\alpha)^{q_i}$. Note that $C_1 = 1$. From (4.15), (4.16) and since $\beta_\mu > q_{\mu+1}$ we get

$$z_{\mu+1}-\alpha = \alpha_\mu-\alpha + z_{\mu+1}-\alpha_\mu = M_{r_\mu} \prod_{i=1}^{\mu} (z_i-\alpha)^{s_i} + o(e_1^{q_{\mu+1}}).$$

Thus

$$(4.23) \quad C_{\mu+1} = M_{r_\mu} \prod_{i=1}^{\mu} C_i^{s_i}.$$

Since $C_1 = 1$ we get after some tedious calculations

$$C_{\mu+1} = M_{r_\mu} \prod_{j=1}^{\mu-1} M_{r_j}^{s_{i+1}(s_{i+2}+1)\dots(s_\mu+1)}$$

which proves the first part of Lemma 4.2.

Let $r_1 > 1$. Assume by induction that $\underline{K}^{q_i-1} \le |C_i| \le \bar{K}^{q_i-1}$. This is true for $i = 1$ since $C_1 = q_1 = 1$. From (4.23) and (4.21) we have

$$|C_{\mu+1}| \le \bar{K}^{r_\mu-1 + s_1(q_1-1) + \dots + s_\mu(q_\mu-1)} = \bar{K}^{q_{\mu+1}-1}$$

and similarly we get a lower bound.

Let $r_1 = 1$. Assume by induction that $c_i \underline{K}^{q_i-1} \le |C_i| \le \bar{K}^{q_i-1} c_i$ where $c_1 = 1$, $c_2 = |M_1|$ and $c_i = |M_1|^{s_2(s_3+1)\dots(s_{i-1}+1)}$ for $i \ge 3$. This is true for $i = 1$ and 2 since $C_1 = q_1 = q_2 = 1$ and $C_2 = M_{r_1}$. Then

$$|C_{\mu+1}| \le \bar{K}^{q_{\mu+1}-1} |M_1|^{s_2+s_2 s_3+s_4 s_2(s_3+1)\dots s_\mu s_2(s_3+1)\dots(s_{\mu-1}+1)} = \bar{K}^{q_{\mu+1}-1} c_{\mu+1}$$

and similarly we get a lower bound. Hence (4.22) holds which completes the proof. ∎

Lemma 4.2 in the case $r_1 > 1$ states that the asymptotic constant C_{k+1} depends exponentially on the order q_{k+1}. This property makes an analysis of the complexity of iteration easier (Traub and Wozniakowski will analyze it in a future paper).

We are now in a position to answer the following question. For given n and k, $k \le n$, find nonnegative integers $s_1, s_2, \ldots, s_k$ to maximize the order of information $p_k = \max_{s_1+\ldots+s_k=n} s_1 \prod_{i=2}^{k} (s_i+1)$. Using a standard technique it is easy to verify that

$$(4.24) \quad \left(n + (k-1)\left\lceil\frac{n-1}{k}\right\rceil\right)\left(1 + \left\lceil\frac{n-1}{k}\right\rceil\right)^{k-1} \le p_k \le \left(\frac{n+k-1}{k}\right)^k < 2^{n-1}$$

for $k \le n-2$ and $p_k = 2^{n-1}$ for $k = n-1$ or n. If k is a divisor of n-1 then the optimal s_i are given by

$$s_1 = 1 + \frac{n-1}{k} \text{ and } s_i = \frac{n-1}{k} \text{ for } i = 2,\ldots,k.$$

For $k = n$ the optimal $s_i \equiv 1$. Furthermore from Theorem 7.1 in Kung and Traub [74] it follows that there are exactly two cases which maximize the order of information,

$$\begin{array}{ll} k = n-1,\ s_1 = 2,\ s_i = 1 & \text{for } i = 2,\ldots,n,\ p_{n-1} = 2^{n-1} \\ k = n,\ s_i = 1 & \text{for } i = 1,\ldots,n,\ p_n = 2^{n-1}. \end{array}$$

The first case means that we use f and f' at the first point and f at the other points. The second case states that we use n function evaluations. From Theorem 4.1 and (4.24) we get

Corollary 4.3

The Kung and Traub conjecture holds for hermitian information ($p_n(0,H) = 2^{n-1}$). ■

The next part of this section deals with the general problem $f^{(m)}(x) = 0$, $m \geq 1$. It seems to us that hermitian information is not always relevant for that problem especially for large m. Note that we have to compute $f^{(m)}(x_1)$ and if the information is hermitian then we have to assume $n \geq m+1$. On the other hand if we use $f^{(m)}(z_1),\ldots,f^{(m)}(z_n)$ (which is nonhermitian) then the order of information is 2^{n-1}. However it is interesting to know the optimal order of information for special hermitian cases, e.g., f, f' at z_1 followed by n-1 function evaluation at the other points for the problem $f'(x) = 0$, (see Lemma 4.5).

Recall that $p_n(m,H)$ denotes the maximal order of hermitian information. In general we do not know $p_n(m,H)$. We only show some bounds on it.

Lemma 4.4

$p_n(m,H) \leq 2^{n-1}$.

Proof

If $\tilde{f} \underset{\mathfrak{N}}{\equiv} f$ then

$$(4.25) \quad \tilde{f}^{(m)}(x) - f^{(m)}(x) = [G(x)\prod_{i=1}^{k}(x-z_i)^{s_i}]^{(m)}$$

for an analytic function G. Let $G(x) = \frac{1}{m!}(x-\alpha)^m$. Since $\tilde{f}^{(m+1)}(\alpha)$ tends to $g^{(m+1)}(\alpha) \neq 0$ as x_1 tends to α then setting $x = \alpha$ in (4.25) we have

$$\tilde{\alpha}-\alpha = c(\alpha,x_1)\prod_{i=1}^{k}(\alpha-z_i)^{s_i}$$

where $c(\alpha,x_1)$ tends to a nonzero limit (see (4.4)).

The proof of Lemma 4.4 may now be obtained analogously to the proof of Theorem 4.1.

Lemma 4.5

Let $n \geq m+1 \geq 2$. Then

$$p_n(m,H) \geq c\, q(m)^{n-1}$$

where

$$c = c(m) = \frac{2}{(1+2m+\sqrt{t})}, \quad q(m) = \left(\frac{1+\sqrt{t}}{2}\right)^{\frac{1}{m}}$$

and $t = 1 + 4m$.

Proof

Define $s_1 = m+1$ and $s_i = m$ for $i = 2,\ldots,k$. Let $z_2 = x_1 + \beta f^{(m)}(x_1)$ for $\beta \neq 0$ and let z_μ, $\mu \geq 3$, be the nearest zero to $z_{\mu-1}$ of the polynomial $w_\mu^{(m)}$ where

$$w_\mu^{(j)}(z_i) = f^{(j)}(z_i), \qquad i = 1,2,\ldots,\mu-1; \quad j = 0,1,\ldots,m-1,$$

$$w_\mu^{(m)}(z_1) = f^{(m)}(z_1)$$

and w_μ is of degree $\leq (\mu-1)m$. It is straightforward to verify that

$$z_\mu - \alpha = O((x_1-\alpha)^{q_\mu})$$

where $q_1 = q_2 = 1$ and for $\mu \geq 3$,

$$q_\mu = m(q_1+\ldots+q_{\mu-2}) + q_1 = q_{\mu-1} + mq_{\mu-2}.$$

It is easy to verify that

$$q_{k+1} \geq c\left(\frac{1+\sqrt{t}}{2}\right)^{k+1}$$

where $c = c(m) = 2/(1 + 2m + \sqrt{t})$.

For a given n let $k = \lfloor (n-1)/m \rfloor = \frac{n-1}{m} + \theta$ where $-1 < \theta \le 0$. The total number of evaluation is equal to $km + 1 \le n$. Hence $p_n(m,H) \ge p_{km+1}(m,H) \ge q_{k+1} \ge$
$1+\theta$
$\ge q_{k+1} \ge c\ q(m)^{n-1} \left(\frac{1+\sqrt{t}}{2}\right) \ge c\ q(m)^{n-1}$ which proves Lemma 4.5. ■

Lemma 4.4 and 4.5 state that $p_n(m,H)$ as a function of n is exponentially bounded from below and above. However $\lim_{m\to\infty} q(m) = 1$.

5. GENERAL INFORMATION, m = 0

We deal with the n-evaluation problem for m = 0. For small n it is possible to verify the Kung and Traub conjecture and to characterize the information sets for all iterations which have maximal order.

For n = 1 the unique piece of information is given by $f(x_1)$. Since $\tilde{f}(x) = f(x) + (x-x_1)$ has the same information as f then $p_1(0) = 1$. This means that for any $y = y(x_1, f(x_1))$ the distance $\alpha - y$ can be at most of first order in $\alpha - x_1$. However y is not, in general, an iteration function, see Lemma 3.3. Note also that for any m, $p_1(m) = 1$.

For n = 2, Kung and Traub [73] proved that the maximal order of iteration equals two under a certain assumption on the iterations considered. Using our technique we find the order of information for any $\mathfrak{N}$ with n = 2. Note that if $\mathfrak{N}$ is hermitian information then $p(\mathfrak{N}) \le 2$, by Corollary 4.3. Thus it suffices to consider the non-hermitian case. Let us first consider one-point iterations, i.e., k = 1 and $\mathfrak{N} = \{f(x_1), f^{(j)}(x_1)\}$ for $j \ge 2$. Then $\tilde{f}(x) = f(x) + (x-x_1)$ and $p(\mathfrak{N}) = 1$. Let us pass to two-point iterations, i.e., k = 2 and $\mathfrak{N} = \{f(x_1), f^{(j)}(z_2)\}$ where $j \ge 1$ and

$z_2 = z_2(x_1, f(x_1))$. If $j \geq 2$ then $\tilde{f}(x) = f(x) + (x-x_1)$ and $p(\mathfrak{N}) = 1$. Let $j = 1$. Then $\tilde{f}(x) = f(x) + (x-x_1)(x-2z_2+x_1)$. From this we get

$$\tilde{\alpha}-\alpha \cong (\alpha-x_1)(\alpha-y), \quad y = 2z_2-x_1.$$

Since $y = y(x_1, f(x_1))$ then $\alpha-y$ can be at most of first order in $(\alpha-x_1)$. Hence $p(\mathfrak{N}) \leq 2$ and $p(\mathfrak{N}) = 2$ if, for instance, $z_2 = x_1 + \beta f(x_1)$, for any constant $\beta \neq 0$.

It is easy to verify that, in addition, $p_2(m) = 2$ for any m.

For $n = 3$, $p_3(0) = 4$. There are a number of information sets $\mathfrak{N}$ for which $p(\mathfrak{N}) = 4$. A proof and discussion may be found in Meersman [75].

Unfortunately the proof technique used to establish the cases $n = 2, 3$ cannot be used for general n since there are too many sub-cases to investigate.

We now wish to discuss some general properties of the n-evaluation problem.

Recall that $E_n^k = (e_{ij})$ is the incidence matrix of the information $\mathfrak{N}$ and let

$$(5.1) \quad M_r = \sum_{j=0}^{r} \sum_{i=1}^{k} e_{ij}$$

denote the total number of evaluations $f, f', \ldots, f^{(r)}$ at $z_1, \ldots, z_k$, $r = 0, 1, \ldots$.

The incidence matrix E_n^k satisfies the <u>Pólya conditions</u> if

$$(5.2) \quad M_r \geq r+1 \qquad \text{for } r = 0, 1, \ldots, n-1.$$

(See Sharma [72].) If E_n^k satisfies the Pólya conditions then $e_{ij} = 0$ for any i and $j \geq n$. This means we do not use derivatives of order higher than n-1. Note that hermitian E_n^k satisfies the Pólya conditions. Furthermore all <u>known</u> information sets with maximal order of information have E_n^k which satisfy the Pólya conditions.

Let $j' = j'(E_n^k)$ be a nonnegative integer such that

$$M_r \geq r+1 \text{ for } r = 0,1,\ldots,j' \text{ and } M_{j'+1} < j'+2.$$

Since $j'+1 \leq M_{j'} \leq M_{j'+1} \leq j'+1$ then $e_{i,j'+1} = 0$ which means that we do not use the (j'+1) derivative. We shall call such $j' = j'(E_n^k)$ <u>an index of E_n^k</u>. E_n^k satisfies the Pólya conditions if and only if its index is equal to n-1.

We introduce the concept of the polynomial order of information pol($\mathfrak{N}$) defined by

$$(5.3) \quad \mathrm{pol}(\mathfrak{N}) = \begin{cases} 0 & \text{if B is empty} \\ \sup B & \text{otherwise} \end{cases}$$

where

$$B = \{q \geq 1 : \forall f \in \mathfrak{F},\ f(\alpha) = 0,\ \forall \tilde{f} \underset{\mathfrak{N}}{=} f \text{ and } \tilde{f}-f \in \Pi_n,$$

$$\limsup_{x_1 \to \alpha} \frac{|\alpha - \tilde{\alpha}|}{|x_1 - \alpha|^{q-\varepsilon}} = 0,\ \forall \varepsilon > 0\},$$

and Π_n denotes a class of polynomials of degree $\leq$ n. Compare with the order of information where is not assumed that $\tilde{f}-f \in \Pi_n$, see (2.10). Thus $p(\mathfrak{N}) \leq \mathrm{pol}(\mathfrak{N})$. Similarly let $\mathrm{pol}(n) = \sup_{\mathfrak{N}} p(\mathfrak{N})$. This gives

$$(5.4) \quad p_n(0) \leq \mathrm{pol}(n).$$

We show some properties of pol(n). From Section 4 it follows that $\mathrm{pol}(n) \geq 2^{n-1}$ and $\mathrm{pol}(n) = 2^{n-1}$ for hermitian information. Furthermore it is possible to show that $pol(n) = 2^{n-1}$ for $n = 1,2,3$ and that pol(n) is an increasing function of n.

Lemma 5.1

Let j' be the index of the incidence matrix E_n^k of $\mathfrak{N}$. Then

$$pol(\mathfrak{N}) \le pol(j'+1).$$

Proof (Compare with the proof of the Schoenberg Lemma in Schoenberg [66] and Sharma [72], Lemma 1.)

Let $E_{j'}^k$ denote the first (j'+1) columns of E_n^k. Assume $f \in \Pi_{j'+1}$. Then $z_i = z_i(x_1; \mathfrak{N}(x_1; f)) = z_i(x_1; \mathfrak{N}_1(x_1; f))$ where $\mathfrak{N}_1$ is the information based on $E_{j'}^k$. Let $h \in \Pi_{j'+1}$ and

$$(5.5) \quad h^{(j)}(z_i) = 0 \quad \text{for } (i,j) \in e_n^k \text{ and } j \le j'.$$

The total number of homogeneous equations in (5.5) is equal to $M_{j'} = j'+1$ and since we have j'+2 unknowns then there exists a nonzero h satisfying (5.5). Furthermore $h^{(j)}(x) \equiv 0$ for $j \ge j'+2$ which means that $h^{(j)}(z_i) = 0$ for all $(i,j) \in e_n^k$. Define $\tilde{f}(x) = f(x) + h(x)$ we get

$$(5.6) \quad \tilde{\alpha}-\alpha = \frac{1}{g'(\alpha)} (1 + o(1))h(\alpha).$$

But $h(\alpha)$ depends only on $E_{j'}^k$ and it can be at most of order pol(j'+1). This proves that $pol(\mathfrak{N}) \le pol(j'+1)$. ■

Since pol(n) is an increasing function of n we immediately have

Corollary 5.2

A necessary condition for $\mathfrak{N}$ to have the maximal polynomial order pol(n) is that its incidence matrix E_n^k satisfies the Pólya conditions. ■

We believe that $\mathrm{pol}(n) = 2^{n-1}$. However to find even a crude upper bound on pol(n) seems to be hard. We give an upper bound on pol(n) under the following conjecture.

Conjecture 5.3

Let $\varphi_1, \varphi_2, \ldots, \varphi_n$ be any n-point iterations. Then there exists a function $f \in \mathfrak{J}$ such that

$$(5.7) \quad \lim_{x_1 \to \alpha} \left| \frac{\varphi_i(x_1;\ \mathfrak{N}(x_1;\ f)) - \alpha}{e_1^{\mathrm{pol}(n)+\varepsilon}} \right| = +\infty, \ \forall \varepsilon > 0, \ \forall i \le n. \quad ■$$

Assume for simplicity that $C_i = C_i(f, \varphi_i) = \lim_{x_1 \to \alpha} |\{\varphi_i(x_1;\ \mathfrak{N}(x_1;\ f)) - \alpha\}/e_1^{\mathrm{pol}(n)}|$ exist for $i = 1, 2, \ldots, n$. The conjecture 5.3 states that they are all different from zero for one function. Note that it holds for $n = 1$.

Lemma 5.4

If (5.7) holds then $\mathrm{pol}(n) < n!$ for $n \ge 3$.

Proof

Let E_n^k be the incidence matrix of $\mathfrak{N}$. Let $0 \not\equiv h \in \Pi_n$ and $h^{(j)}(z_i) = 0$ for $(i,j) \in e_n^k$. Then

$$h(x;x_1) = a(x_1)(x-h_1)(x-h_2)\ldots(x-h_j)$$

where $1 \le j \le n$ and $a(x_1)$ is chosen in order to ensure that $h(x;x_1)$ tends to an analytic function as x_1 tends to α. Note that $h_1 = x_1$ and $h_i = h_i(z_1, z_2, \ldots, z_k)$ depends on at most (n-1) evaluations. If $\lim_{x_1 \to \alpha} h_i = \alpha$ then h_i can be treated as an iteration. From (5.7) we get

$$|h_i - \alpha| \ge c|e_1|^{\mathrm{pol}(n-1)+1-\varepsilon}, \ c > 0,$$

for any $\epsilon > 0$. Since it holds for any $\mathfrak{N}$ we have

$$\text{pol}(n) \leq (n-1)\ \text{pol}(n-1) + 1 < n\ \text{pol}(n-1) \leq n! \qquad \blacksquare$$

The next part of this section deals with a restrictive class of n-point iterations. We use n evaluations per step and we assume that an iteration is exact for a function $f \in \Pi_{n-1}$. We shall say that $\underline{\varphi \in \Phi_n}$ if $\varphi(x_1; \mathfrak{N}(x_1; f)) = \alpha$ whenever $f \in \Pi_{n-1}$ and x_1 is close to α. Note that all iterations considered in Section 4 belong to Φ_n.

Next we shall say that the problem is <u>locally well-poised</u> for f if for every $h \in \Pi_{n-1}$ such that

$$h^{(j)}(z_i) = 0 \quad \text{for} \quad (i,j) \in e_n^k$$

it follows $h \equiv 0$ for all x_1 close to x.

Note that Birkhoff interpolation for E_n^k is well-poised if $\forall (x_1, x_2, \ldots, x_k)\ h^{(j)}(z_i) = 0$ for $(i,j) \in e_n^k$ and $h \in \Pi_{n-1} \Rightarrow h \equiv 0$ (see Sharma [72]). Thus, if Birkhoff interpolation is well-poised than the problem is locally well-poised but not in general vice versa.

<u>Lemma 5.5</u>

If an iteration φ is exact for $f \in \Pi_{n-1}$, $\varphi \in \Phi_n$, then

(i) E_n^k satisfies the Pólya conditions,

(ii) the problem is locally well-poised for $f \in \Pi_{n-1}$,

(iii) $p(\mathfrak{N}) \leq n(n+1)^{n-1}$.

<u>Proof</u>

Suppose that the problem is not locally well-poised for $f \in \Pi_{n-1}$. Then there exists a nonzero $h \in \Pi_{n-1}$ such that

$h^{(j)}(z_i) = 0$ for $(i,j) \in e_n^k$. Define $\tilde{f}(x) = f(x) + h(x)$. Since $\tilde{f} \in \Pi_{n-1}$ and $\tilde{f}(\alpha) \neq 0$ then

$$\alpha = \varphi(x_1, \mathfrak{N}(x_1,f)) = \varphi(x_1, \mathfrak{N}(x_1,\tilde{f})) \neq \tilde{\alpha}.$$

This contradicts that $\varphi \in \Phi_n$. Hence (ii) holds. Let j' be the index of E_n^k. If $j' < n-1$ then there exists a nonzero $h \in \Pi_{j'+1}$ such that $h^{(j)}(z_i) = 0$ for all $(i,j) \in e_n^k$, see the proof of Lemma (5.1). This contradicts that the problem is locally well-poised. Thus, (i) holds.

To prove (iii) it suffices to note that if

$$E_n^k \le \tilde{E}_{\tilde{n}}^k \quad \text{then} \quad p(E_n^k) \le p(\tilde{E}_{\tilde{n}}^k)$$

for $n \le \tilde{n}$ where by $E_n^k = (e_{ij}) \le \tilde{E}_{\tilde{n}}^k = (\tilde{e}_{ij})$ we mean $e_{ij} \le \tilde{e}_{ij}$ for $(i,j) \in e_n^k$.

Define $\tilde{E}_{\tilde{n}}^k$ as a hermitian matrix where $\tilde{n} = kn$,

$$\tilde{e}_{ij} = 1 \quad \text{for } i = 1,2,\ldots,k \text{ and } j = 0,1,\ldots,n-1.$$

Of course $E_n^k \le \tilde{E}_{\tilde{n}}^k$ and from Theorem 4.1 we get

$$p(\tilde{E}_{\tilde{n}}^k) \le n(n+1)^{n-1}$$

which proves (iii). ∎

6. FINAL REMARKS

The problem of the maximal order of n-point iterations is connected with Birkhoff interpolation which has been open almost 70 years. The main difficulty is to estimate the difference between the zeros, $\tilde{\alpha}-\alpha$, of any two functions with the same information, $\tilde{f} \underset{\mathfrak{N}}{\equiv} f$. Note that $\tilde{f}$ can belong to Π_{n-1} for all f if the problem is well-poised. However up to now we do not know when Birkhoff interpolation is well-poised. There are many reasons to believe that hermitian information (interpolation without gaps) is optimal. However there also exists nonhermitian information with order 2^{n-1}.

For nonhermitian information $\mathfrak{N}$ it is hard to find the order $p(\mathfrak{N})$. We know the order of such information only in a few cases. The first one is a Brent iteration based on $\mathfrak{N} = \{f(z_1), f'(z_1), \ldots, f^{(j)}(z_1), f^{(r)}(z_2), f^{(r)}(z_3), \ldots, f^{(r)}(z_k)\}$ for suitable chosen z_i where $0 < r \le j+1$ (see Brent [75]). This information uses $n = j+k$ evaluations and has the order $p(\mathfrak{N}) = j + 2k - 1$, see Meersman [75]. Note that this problem is well-poised. The second example is Abel-Goncarov information given by

$$\mathfrak{N} = \{f(z_1), f'(z_2), \ldots, f^{(n-1)}(z_n)\},$$

see Sharma [72]. Recall that if $z_i = z_1$ for $i = 2, \ldots, n$ then we get one-point information which has the order n (even in the multivariate and abstract cases). For Abel-Goncarov information it is possible to prove

$$n \le p(\mathfrak{N}) \le 2n$$

but we do not know whether this upper bound is sharp. Finally let us mention lacunary information given by

$$\mathfrak{N} = \{f(z_1), f''(z_1), f(z_2), f''(z_2), \ldots, f(z_k), f''(z_k)\}$$

and $n = 2k$, see Sharma [72]. It is possible to verify that

$$\frac{1}{2}\, 2^{n/2} \le p(\mathfrak{N}) \le \frac{3}{4}\, 2^n$$

but the exact value of $p(\mathfrak{N})$ is unknown.

ACKNOWLEDGMENT

I am greatly appreciative to J. F. Traub, A. Sharma, B. Kacewicz and R. Meersman for their helpful comments and assistance during the preparation of this paper.

REFERENCES

Brent [75] Brent, R. P.,"A Class of Optimal Order Zero Finding Methods Using Derivative Evaluation," Department of Computer Science Report, Carnegie-Mellon University, 1975, these Proceedings.

Kung and Traub [73] Kung, H. T. and J. F. Traub, "Optimal Order for Iterations Using Two Evaluations," Department of Computer Science Report, Carnegie-Mellon University, 1973. To appear in SIAM J. Numer. Anal.

Kung and Traub [74] Kung, H. T. and J. F. Traub, "Optimal Order of One-Point and Multipoint Iterations," J. Assoc. Comput. Mach., Vol. 21, No. 4, 1974, 643-651.

Meersman [75] Meersman, R., "Optimal Use of Information in Certain Iterative Processes," Department of Computer Science Report, Carnegie-Mellon University, 1975, these Proceedings.

Traub [61] Traub, J. F., "On Functional Iteration and the Calculation of Roots," Proc. 16th Nat. ACM Conf., 5A-1 (1961), 1-4.

Traub [64]	Traub, J. F., Iterative Methods for the Solution of Equations, Prentice-Hall, Englewood Cliffs, N. J., 1964.
Schoenberg [66]	Schoenberg, I, J., "On Hermite-Birkhoff Interpolation," J. Math. Anal. Appl., 16 (1966), 538-543.
Sharma [72]	Sharma, A., "Some Poised and Nonpoised Problems of Interpolation," SIAM Review, Vol. 14, No. 1, 1972, 129-151.
Woźniakowski [75]	Woźniakowski, H., "Generalized Information and Maximal Order of Iteration for Operator Equations," SIAM J. Numer. Anal., Vol. 12, No. 1, 1975, 121-135.

OPTIMAL USE OF INFORMATION IN CERTAIN ITERATIVE PROCESSES

Robert MEERSMAN
Vrije Universiteit Brussel

1. Introduction

To approximate numerically a zero α of a real analytic function f,

$$f(\alpha) = 0,$$

iteration is most widely used. We will discuss so-called *k-point stationary iterative processes without memory*, ϕ, defined as follows. Let there exist an interval around α such that for all x_1 in this interval, the following functions ζ_i, i=2,...,k+1 are well-defined :

$$z_1 = x_1$$
$$z_2 = \zeta_2(z_1, N(z_1;f))$$
$$z_3 = \zeta_3(z_1, z_2, N(z_1, z_2;f))$$
$$\vdots$$
$$x_2 := \phi(x_1) = z_{k+1} = \zeta_{k+1}(z_1,\ldots,z_k, N(z_1,\ldots,z_k;f)),$$

where

a) x_2 is called the new approximation to α ; we assume that x_2 lies within the same interval a and that the process is converging, i.e. putting $x_1 := x_2$ and repeating the iterative step produces a sequence which converges to α. Also we assume $f'(\alpha) \neq 0$.

b) $N(z_1,\ldots,z_i;f)$ is called the *information set* at

the points $z_1,\ldots,z_i$ and consists of the set of all evaluations of f and/or its derivatives used at those points when computing z_{i+1}. It is not necessary to give *all* derivatives up to a certain degree at any particular point z_i. Examples of specific N abound in the rest of the paper. Other kinds of information sets can be considered, but are not the topic of this paper ; see for instance Kacewicz [75]. The reader should also be more or less familiar with the work of Woźniakowski [74] [75a].

2. Some known results and conjectures

Definition 2.1.

Let $N(z_1,\ldots,z_k;f)$ be given. We say that $\tilde{f} \equiv f \bmod N$ when for all $f^{(j)}(z_i) \in N$,

$$f^{(j)}(z_i) = \tilde{f}^{(j)}(z_i)$$

As in Woźniakowski [75a] we adopt the following definitions:

Definition 2.2.

The order of iteration $p(\phi)$ is the largest number such that for all f with $f(\alpha) = 0$, for all $\tilde{f} \equiv f \bmod N$ and having a zero $\tilde{\alpha}$ near α, we have

$$\limsup_{x_1 \to \alpha} \frac{|\phi(x_1;N) - \tilde{\alpha}|}{|x_1 - \alpha|^{p(\phi)}} < \infty$$

It can be shown this definition coincides with the usual definition in the literature (e.g. Traub[64]) under weak assumptions on the asymptotical constant.

Definition 2.3.

The order of information $p(N)$ is the largest number such that for all f with $f(\alpha) = 0$, for all $\tilde{f} \equiv f \bmod N$ and having a zero $\tilde{\alpha}$ near α, we have

$$\limsup_{x_1 \to \alpha} \frac{|\alpha - \tilde{\alpha}|}{|x_1 - \alpha|^{p(N)}} < \infty$$

Remark that choosing a special f or $\tilde{f} \equiv f \bmod N$ will give upper bounds on these orders, by definition. See also Woźniakowski [75b]. Extensive use will be made of the following two theorems.

Theorem 2.1. (Maximal Order Theorem)

The order $p(\phi)$ is bounded above by $p(N)$ for all ϕ using information N.

Theorem 2.2.

The maximal order is reached for the generalized interpolating methods $I_N : p(I_N) = p(N)$.

For definitions and proofs, see Woźniakowski [75a].

Two kinds of problems arise now.

Problem 1. Given N, compute $p(N)$. (In other words, what is the maximal order achievable with information N).

Problem 2. Given $n = \# N$, the number of elements in N, determine $P_n^* = \max_{\# N = n} \max_{\phi} p(\phi)$, where ϕ uses information N.

Problem 2 is much harder than problem 1. Kung and

Traub conjectured (Kung and Traub [74a]) that

(2.3) $P_n^* = 2^{n-1}$

and exhibited two families of methods which realize this bound for each n. In a later paper, Kung and Traub [74b], they proved (2.3) for n = 1 and 2. Remark that in view of Theorems 2.1 and 2.2, (2.3) is equivalent to

$$Q_n^* = 2^{n-1}$$

where $Q_n^* = \max_{\#N=n} p(N)$.

The conjecture has been settled by Woźniakowski in one very important case.

Definition 2.4.

N is *hermitean* iff for all i, $1 \leq i \leq k$, we have

$$f^{(k)}(z_i) \in N \Rightarrow f^{(k-1)}(z_i) \in N \text{ for all } k > 0.$$

Theorem 2.3. (Woźniakowski [75b])

The conjecture of (2.3) is true if the maximum is taken over hermitean N. We will show later that a partial converse is not true, i.e. that $p(N) = 2^{\#N-1}$ does not imply that N is hermitean.

3. A solution to problem 1 for n = 3

We will prove in this section that $P_3^* = 4$, showing the correctness of the conjecture in this case. Our proof uses special cases of some general results on certain n-evaluations iterations one of which is to be treated in a later paper, namely

Theorem 3.1.

If $N = \{f(z_1), f^1(z_2), \ldots, f^{(n)}(z_{n+1})\}$ then

$$p(N) \leqslant 2n$$

Note : N is the so-called Abel-Contarov information.

In the proof of the following lemma and the rest of the paper we assume that at each z_i used in ϕ, some new information is computed. This is not a restriction since otherwise we can substitute the expressions for these z_i in the other ζ_j, obtaining an equivalent iteration (with less points).

Lemma 3.1.

Let ϕ be an iteration using two pieces of information, i.e.

$$z_1 = x_1$$

$$z_2 = \zeta_2(z_1, N(z_1;f))$$

$$x_2 := \phi(x_1) = z_3 = \zeta_3(z_1, z_2, N(z_1, z_2; f)) \text{ with } \#N = 2$$

(By the above convention, ϕ is a 1- or 2-point iteration).

Then, if there exists a (known) constant C, $C \neq 0$ and $C \neq 1$, such that for all f

$$\alpha - z_2 = C(\alpha - z_1) + O(\alpha - z_1)^2, \qquad (3.1)$$

then ϕ cannot be of second order.

Proof : If $C \neq 1$, then from (3.1) we could solve for α :

$$\alpha = \frac{z_2 - C\, z_1}{1 - C} + O(\alpha - z_1)^2$$

And $z_2^* = \frac{z_2 - C z_1}{1 - C}$ would therefore produce a second order approximation to α. Since $P_1^* = 1$, both pieces of information must be used then at z_1, but then ϕ is a one-point iteration, i.e. $z_3 = z_2$ by the convention, and from (3.1) and $C \neq 0$ it follows that ϕ is only of first order.

Theorem 3.2.

$$P_3^* = 4$$

Proof : We prove that $p^*(N) \leq 4$ for all N with $\#N = 3$, where

$$p^*(N)=\max \{p \mid \text{for all } \tilde{f}=f+G,\ G\equiv 0 \bmod N,\ \tilde{f}(\tilde{\alpha})=0, \text{ and } G \text{ monic polynomial of degree} < 3,\ \limsup_{x_1\to\alpha} \frac{|\alpha-\tilde{\alpha}|}{|x_1-\tilde{\alpha}|^p} < \infty\}$$

thereby restricting the class of $\tilde{f}$ such that $\tilde{f} \equiv f \bmod N$.

Step 1

We need one evaluation of f at z_1 to assure convergence so N is of the form

$$N = \{f(z_1),\ f^{(i)}(z_2),\ f^{(j)}(z_3)\}$$

(Kung and Traub [74a]).

We will suppose z_2 and z_3 not necessarily different, unless of course i or j equals 0 or i = j. It is clear this does not affect the bounds on the optimal order.

Since now G is a monic polynomial of degree < 3, we can take i and j $\leq$ 2. Indeed, if i or j > 3, $G \equiv 0 \bmod N$ is automatically satisfied ; if

$i < j = 3$, we can interpolate the zero function for this information at z_1 and z_2 with a monic polynomial of degree ≤ 2 - from $P_2^* = 2$ it follows that the optimal order is $2 < 4$, and similarly if $j < i = 3$. If $i = j = 3$ we can even take $G(z) = z - z_1$ in which case the order of information evidently is equal to $1 < 4$.

<u>Remark</u>

The above argument can of course be generalized to any n : it is closely related to the Pólya conditions on the set N, see Woźniakowski [75b], Sharma [72].

<u>Step 2</u>

The different cases for the information N.

With an obvious notation, in the following cases the answer is already known :

<u>Case 1</u> : $\{f_1, f_2, f_3\}$: Hermitean N, order $\leq 2^{3-1} = 4$

<u>Case 2</u> : $\{f_1, f_2', f_3'\}$: "Brent information with m=0", applying the results of Sec. 4, we find

if $z_1 \neq z_2 : p(N) \leq m+2(k-1)-1 = 0+4-1 = 3 < 4$;

if $z_1 = z_2 : p(N) = m+2(k-1)+1 = 1+2+1 = 4$

<u>Case 3</u> : $\{f_1, f_2', f_3''\}$: "Abel-Goncarov" N, by Theorem 3.1. : order $\leq 2.2 = 4$

<u>Case 4</u> : $\{f_1, f_2'', f_3''\}$: Take again $G(z) = z - z_1$ as in Step 1 ; order $\leq 1 < 4$ (here the Pólya conditions are not satisfied).

Step 3

Exhaustive checking of the remaining cases. Let us set

$$G(z) = (z-z_1)(z^2+az+b)$$

Case 5 : $\{f_1,f_2,f_3'\}$

Now $G(z) = (z-z_1)(z-z_2)(z-c)$.

The condition $\tilde{f}'(z_3) = 0$ gives

$$(2z_3-z_1-z_2)(z_3-c)+(z_3-z_1)(z_3-z_2) = 0$$

So in general, c is a function of z_1, z_2 and z_3 which itself is also a function of z_1 and z_2. Since $P_2^* = 2$, $\alpha-c$ cannot be of higher order than $(\alpha-z_1)^2$. Therefore

$$\tilde{\alpha} - \alpha = \mathcal{O}(G(\alpha)) = \mathcal{O}(\tilde{f}(\alpha)) = \mathcal{O}(\alpha-z_1)(\alpha-z_2)(\alpha-c)$$

cannot be of higher order than $(\alpha-z_1)^4$ since $\alpha-z_2$ is at most of order $(\alpha-z_1)$ because $P_1^* = 1$. Thus $p^*(N) \leq 4$ for this N.

Case 6 : $\{f_1,f_2,f_3''\}$

Completely analogous to case 5, the condition at z_3 now is

$$(z_3-c) + (2z_3-z_1-z_2) = 0.$$

Case 7 : $\{f_1,f_2'',f_3\}$

Now $G(z) = (z-z_1)(z-z_3)(z-c)$.

The condition at z_2 : $c = 3z_2 - z_1 - z_3$, again gives that $(\alpha-c)$ is at most of second order in $(\alpha-z_1)$.

Now $(\alpha-z_3)$ is at most of first order in $(\alpha-z_1)$ since the function

$$\tilde{G}(z) = z - z_1$$

interpolates the zero function at the points z_1 and z_2 for the given information. Thus with an obvious notation,

$\tilde{\alpha}(\tilde{G}) - \alpha = \mathcal{O}(\alpha-z_1)$, and by theorem 2.1 and 2.2,

$$(\alpha-z_3) = \mathcal{O}(\alpha-z_1) \text{ at most.}$$

Consequently,

$$\tilde{\alpha}(G)-\alpha = \mathcal{O}(\tilde{f}(\alpha))=\mathcal{O}(\alpha-z_1)(\alpha-z_3)(\alpha-c)$$

is again at most of order $(\alpha-z_1)^4$.

Case 8 : $\{f_1, f_2'', f_3'\}$

This is a permutation of the Abel-Gončarov information. It is an easy consequence of the proof of Theorem 3.1 that also here we have,

$$p(N) \leq 2 - 2 = 4.$$

A direct proof for this case can also be found, and is left to the reader.

Case 9 : $\{f_1, f_2', f_3\}$

$G(z) = (z-z_1)(z-z_3)(z-c)$.

Again, $(\alpha-c)$ being of 3rd order or more in $(\alpha-z_1)$ would contradict $P_2^* = 2$, so if $(\alpha-z_3)$ is of order 1 in $(\alpha-z_1)$ we are done. However $(\alpha-z_3)$ and $(\alpha-c)$ cannot be both of second order, since the condition $G'(z_2) = 0$ reads

$$(2z_2-z_1-z_3)(z_2-c)+(z_2-z_1)(z_2-z_3) = 0$$

which is easily seen to be equivalent to

$$e_3(\alpha-c)=e_2(2e_1-3e_2)-(e_1-2e_2)(e_3+(\alpha-c))$$

where $e_i = \alpha - z_i$; $i=1,2,3$.

Now $e_3(\alpha-c)$ cannot be of order 4 while by lemma 3.1 with $C = \frac{2}{3}$, $e_2(2e_1-3e_2)$ is at most of order 2 and $(e_1-2e_2)(e_3+(\alpha-c))$ is at least of third order. Note that if $2z_2 - z_1 - z_3 = 0$, this case becomes non-poised (Sharma [72]) and the function $G(z) = (z-z_1)(z-z_3)$ interpolates zero with respect to this N, giving a maximal order of 3.

Theorem 3.3.

Hermitean information is not uniquely optimal.

Proof : We exhibit two examples for $n = 3$

a) Consider again case (6) of Theorem 3.2.

We have $G(\alpha) = (\alpha-z_1)(\alpha-z_2)(\alpha-c)$

where $c = 3z_3 - z_1 - z_2$.

To obtain order 4, this suggests we must have

$$\alpha - c = O(\alpha-z_1)^2, \text{ or}$$

(3.1) $$z_3 = \frac{z_1+z_2}{3} + \frac{1}{3}\gamma \text{ with } \gamma-\alpha=\theta(\alpha-z_1)^2$$

To find such a γ, take

$$z_1 = x_1$$

$$z_2 = z_1 + f(z_1)$$

and γ the root of the interpolating (1st degree) polynomial at these two points, i.e.

$$\gamma = z_1 - \frac{z_2 - z_1}{f(z_1) - f(z_1)} f(z_1)$$

Then construct z_3 by (3.1) and

$$x_2 := \phi(x_1) := z_4$$

as the root of the (2nd degree) polynomial interpolating f with respect to the information at z_1, z_2 and z_3. This method is easily seen to be of fourth order.

b) Case 3 with $z_1 = z_2$. Now $G(z) = (z-z_1)^2(z-c)$ and $G''(z_3) = 0$ gives $3z_3 = 2z_1 + c$. By taking $z_3 = \frac{2}{3}z_1 + \frac{1}{3}\gamma$ where again $\alpha - \gamma = O(\alpha - z_1)^2$, for example by a Newton-step, it is easy to show that the following method has fourth order :

$$z_1 = x_1 \quad (= z_2)$$
$$z_3 = \frac{2}{3} z_1 + \frac{1}{3} \left(z_1 - \frac{f(z_1)}{f'(z_1)}\right)$$

$x_2 := \phi(x_1) := z_4$ = zero of the second degree polynomial interpolating f with respect to the information at z_1 and z_3.

Remarks

The previous arguments permit the determination of all arrangements of the information ($\# N = 3$) which can give optimal order. They are denoted by their incidence matrices (Sharma [72]) as follows :

A) $\begin{pmatrix} 1 & 1 \\ 1 & 0 \end{pmatrix}$ B) $\begin{pmatrix} 1 & 1 \\ 0 & 1 \end{pmatrix}$ C) $\begin{pmatrix} 1 & 0 & 0 \\ 1 & 0 & 0 \\ 1 & 0 & 0 \end{pmatrix}$ D) $\begin{pmatrix} 1 & 1 & 0 \\ 0 & 0 & 1 \end{pmatrix}$

E) $\begin{pmatrix} 1 & 0 & 0 \\ 1 & 0 & 0 \\ 0 & 0 & 1 \end{pmatrix}$ F) $\begin{pmatrix} 1 & 0 \\ 1 & 0 \\ 0 & 1 \end{pmatrix}$ G) $\begin{pmatrix} 1 & 0 & 0 \\ 0 & 1 & 0 \\ 0 & 0 & 1 \end{pmatrix}$

The cases A) and C) have optimal generalizations for all $n > 3$, it will be shown in a later paper that also case F) can be generalized to a *non-hermitean* optimal case for all n.

4. A solution to a "problem 2" for N = Brent information.

Definition 4.1.

$$N_{m,\ell,k}=\{f(z_1),f'(z_1),\ldots,f^{(m)}(z_1)\ ;\ f^{(\ell)}(z_2),\ldots,f^{(\ell)}(z_k)\}$$

is called Brent-information, where the z_i are distinct, $m \geq 1$, $k \geq 2$ and $\ell \geq 1$.

Brent has shown the following

Theorem 4.1. (Brent [74])

Assume $\ell \leq m + 1$.

There exist methods using $N_{m,\ell,k}$ of order $m+2k-1$.

We will now prove that the Brent methods make optimal use of the information $N_{m,\ell,k}$ (with respect to order).

Theorem 4.2.

Let $N_{m,\ell,k}$ be as in definition 4.1. Then if $\ell \leq m+1$,

$$p(N) = m + 2k - 1$$

If $\ell > m + 1$, $p(N) = m + 1$.

Proof : A technique is used similar to theorem 3.2. If $\ell > m + 1$, a function $G(z) \equiv 0 \bmod N$ is given by

$$G(z) = (z-z_1)^{m+1}$$

And consequently $|\tilde{\alpha}-\alpha| = O(\tilde{f}(\alpha)) = O(\alpha-z_1)^{m+1}$.

By theorem 2.1 the maximal order is not more than $m + 1$. Methods realizing this order exist and are trivial to find. Thus $p(N) = m + 1$.

If $\ell \leq m + 1$, we construct $G(z)$ as follows.

To satisfy the conditions at $z_2, \dots, z_k$ we must have

$$G^{(\ell)}(z) = (z-z_1)^{m-\ell+1}(z-z_2)(z-z_3)\dots(z-z_k)H(z)$$

where $H(z)$ is any (sufficiently regular) function.
Integrating ℓ times,

$$G(z) = \int_{z_1}^{z} (t-z)^{\ell-1}(t-z_1)^{m-\ell+1}(t-z_2)\dots(t-z_k)H(t)\,dt.$$

According to the remark at (2.2), we obtain an upper bound by choosing a special H.
Take $H(t) = (t-z_2)\dots(t-z_k)$, making

$$G(z) = \int_{z_1}^{z} (t-z)^{\ell-1}(t-z_1)^{m-\ell+1}[(t-z_2)\dots(t-z_k)]^2\,dt.$$

Consider now $G(\alpha) = \tilde{f}(\alpha)$. Transform $G(\alpha)$ to the interval $[-1,+1]$; after some easy calculations we obtain

$$\tilde{f}(\alpha) = K(\alpha-z_1)^{\ell-1}(\alpha-z_1)^{m-\ell+1}(\alpha-z_1)^{2(k-1)}(\alpha-z_1).I(\alpha)$$

where $K \neq 0$ does not depend on α or any of the z_i and where

$$I(\alpha) = \int_{-1}^{+1} (1-\tau)^{\ell-1}(1+\tau)^{m-\ell+1} \prod_{i=2}^{k} \left(\tau+\frac{z_1-z_i}{\alpha-z_1}\right)^2 \alpha\,d\tau.$$

As is well known, $I(\alpha)$ is minimized when

$\prod_{i=2}^{k} \left(\tau + \frac{z_1-z_i}{\alpha-z_1}\right)$ is equal to the $(k-1)^{st}$ monic Jacobi polynomial corresponding to the weight function $(1-\tau)^{\ell-1}(1+\tau)^{m-\ell+1}$.

Then $I(\alpha) \geqslant c$ where c is independent of the z_i.
(See for example G. Natanson : "Konstruktive Funktionentheorie").

So $|\tilde{\alpha}-\alpha| = O(\tilde{f}(\alpha)) = O(\alpha-z_1)^p$

with $p = (\ell-1) + (m-\ell+1) + 2(k-1) + 1$
$= m + 2k - 1.$

Thus $p(N) \leq m + 2k - 1$, but by Brent's theorem and theorem 2.1, we have equality.

Note : The previous theorem was independently discovered by Woźniakowski. A generalization of this theorem is possible.

Theorem 4.3.

Let now $N = \{f(z_1), f'(z_1), \ldots, f^{(m_1)}(z_1)$;
$f^{(\ell)}(z_2), \ldots, f^{(\ell+m_2)}(z_2)$;
$f^{(\ell)}(z_3), \ldots, f^{(\ell+m_3)}(z_3)$;
...
$f^{(\ell)}(z_k), \ldots, f^{(\ell+m_k)}(z_k)\}$ $m_1 \geq 1$,
$\ell \geq 1$.

Then if $\ell > m_1 + 1$, $p(N) = m_1 + 1$, and if $\ell \leq m_1+1$,

$$p(N) \leq 1 + m_1 + 2.\sum_{i=1}^{k}\left[\frac{m_i+1}{2}\right] \tag{4.1}$$

Proof : The proof runs analogously, with H replaced by

$$H(t) = \prod_{i=2}^{k}(t-z_1)^{\varepsilon_i} \text{ with } \varepsilon_i = \begin{cases} 0 \text{ if } m_i \text{ odd} \\ 1 \text{ if } m_i \text{ even} \end{cases}$$

(The Jacobi polynomial is now of degree $\sum_{i=2}^{k}\left[\frac{m_{i+1}}{2}\right]$)

Remark 4.1.

Let $m_i = 1$ for $i = 2,\ldots,k$. Then $p(N) \leq m + 2k - 1$, so we gain nothing compared to the Brent information case ! In general the order cannot be raised if all m_i are even and we add the pieces of information $f^{(\ell+m_i+1)}(z_i)$ for $i = 2,\ldots,k$.

Remark 4.2.

Contrary to Theorem 4.2 the inequality (4.1) is not yet known to be an equality in general. If all m_i are equal, methods can be constructed by means of so-called "s-polynomials" realizing the bound. For a definition of these, see Ghizetti and Ossicini, "Quadrature Formulae". Again, for details we refer to a forthcoming paper on this subject.

Remark 4.3.

Kung and Traub's conjecture states that the optimal order for a given number of pieces of information will double by adding one extra piece of information. That it is however possible to increase the order more than twofold when it is not optimal, is shown by the following example :

Let $N = \{f(z_1), f'(z_1), f''(z_1)\ ;\ f'(z_2), f''(z_2)\}$.

By theorem 4.3 and remark 4.1 any method using this information must have order at most 5. Adding the element $\{f(z_2)\}$ to N, we get however an information at which allows us to obtain order 12, as is easily shown. Although of course not a counterexample to the conjecture - we believe it is true - it will complicate any possible proof by induction.

Finally, we state without proof the following result, used in the proof of Theorem 3.1 :

Theorem 4.4.

If in Theorem 4.3 $m_1 = 0$ (and consequently, to avoid trivial cases, $l = 1$) the order of information is bounded by

$$p(N) \leq 1 + m_2 + \sum_{i=3}^{k} \left\lceil \frac{m_i+1}{2} \right\rceil .$$

References

Brent [74] Brent, R., "Efficient Methods for Finding Zeroes of Functions whose Derivatives are Easy to Evaluate," Department of Computer Science Report, Carnegie-Mellon University, December, 1974.

Kacewicz [75] Kacewicz, B., "An Integral-interpolatory Methof for the Solution of Non-linear Scalar Equations," Department of Computer Science Report, Carnegie-Mellon University, January, 1975.

Kung and Traub [74a] Kung, H. T. and Traub, J. F., "Optimal Order of One-point and Multipoint Iteration," JACM 21, 643-651.

Kung and Traub [74b] Kung, H. T. and Traub, J. F., "Optimal Order and Efficiency for Iterations with Two Evaluations," to appear in SIAM J. Numer. Anal.

Sharma [72] Sharma, A., "Some Poised and Non-poised Problems of Interpolation," SIAM Review 14, 1972.

Traub [64] Traub, J. F., Iterative Methods for the Solution of Equations, Prentice-Hall, 1964.

Woźniakowski [74] Woźniakowski, H., "Maximal Stationary Iterative Methods for the Solution of Operator Equations," SIAM J. Numer. Anal. 11, 1974, 934-949.

Woźniakowski [75a] Woźniakowski, H., "Generalized Information and Maximal Order of Iteration for Operator Equations," SIAM J. Numer. Anal. 12, 1975, 121-135.

Woźniakowski [75b] Woźniakowski, H., "Maximal Order of Multipoint Iterations Using N Evaluations," these Proceedings.

THE USE OF INTEGRALS IN THE SOLUTION OF NONLINEAR EQUATIONS IN N DIMENSIONS

B. Kacewicz
Department of Computer Science
Carnegie-Mellon University
(On leave from University of Warsaw)

ABSTRACT

We introduce the maximal order iteration $I_{-1,s}$ for solving of the nonlinear equation $F(x) = 0$ in the N dimensional Banach space, $1 \le N \le +\infty$, which uses the "integral information". Integral information consists of the "standard information" $F^{(j)}(x_d)$, $j=0,1,\ldots,s$ and the value of $\int_0^1 F(x_d + ty_d)dt$ where $s \ge 1$, x_d is close to the solution and y_d only depends on the standard information. We show $I_{-1,s}$ is of order $s + 3 - \delta$, where $\delta = 0$ for $N = 1$ or $s \ge 2$ and $\delta = 1$ otherwise. Since the maximal order for the standard information is equal to $s+1$, the additional value of the integral which is represented by a vector of size N increases the order by $2 - \delta$.

1. INTRODUCTION

We consider the problem of solving the nonlinear equation

(1.1) $F(x) = 0,$

where $F: D \to B_2$, D is an open convex subset of B_1 and B_1, B_2 are Banach spaces, $\dim(B_1) = \dim(B_2) = N$, $1 \le N \le +\infty$. We usually solve (1.1) by iteration and it is often assumed we know the standard information for F, i.e.,

$$\mathfrak{N}_s = \mathfrak{N}_s(x_d;F) = \{F(x_d), F'(x_d), \ldots, F^{(s)}(x_d)\},$$

where $s \ge 1$ and x_d is an approximation to the solution α. In a previous paper (Kacewicz [75]) we raised the question how other types of information can be used in iteration and what is the maximal order of convergence for such information. We answered this question in the case of "integral information" for scalar equations, N = 1. This paper deals with integral information in the multivariate and abstract case.

By integral information we mean

$$(1.2) \quad \mathfrak{N}_{-1,s} = \mathfrak{N}_{-1,s}(x_d;F) = \{F(x_d), F'(x_d), \ldots, F^{(s)}(x_d), \int_0^1 F(x_d + ty_d)dt\},$$

where y_d depends on $x_d, F(x_d), \ldots, F^{(s)}(x_d)$ and $s \ge 1$.

Note that $\mathfrak{N}_{-1,s}$ differs from $\mathfrak{N}_s$ by the additional value of an integral which is represented for $N < +\infty$ by a vector of size N.

In Section 2 we define the iteration $I_{-1,s}$ which uses information $\mathfrak{N}_{-1,s}$ and is of order $s + 3 - \delta$, where

$$\delta = \begin{cases} 0 \text{ if } s \ge 2 \text{ or } N = 1 \\ 1 \text{ if } s = 1 \text{ and } N \ge 2 \end{cases}$$

A theorem on the convergence of $I_{-1,s}$ is proved. The main result of this paper is that for optimally chosen y_d the maximal order of iteration is equal to $s + 3 - \delta$, i.e., that $I_{-1,s}$ has order as high as possible. Since the maximal order of the standard information $\mathfrak{N}_s$ is equal to $s + 1$ the additional use of the integral (i.e. N new inputs) increases the maximal order by $2 - \delta$. Note that we can attain the order $s + 3 - \delta$ by using the standard information $\mathfrak{N}_{s+2-\delta}$ which is expressed by $O(N^{s+3-\delta})$ scalar function evaluations whereas the same order is achievable by $\mathfrak{N}_{-1,s}$ using $O(N^{s+1})$ scalar function evaluations. For example, for $2 \le N < +\infty$ using integral information $\mathfrak{N}_{-1,1}$, i.e., $O(N^2)$ scalar function evaluations we have order 3 whereas the same order is achievable by $\mathfrak{N}_2$, i.e., by $O(N^3)$ scalar function evaluations. Informations $\mathfrak{N}_{-1,2}$ and $\mathfrak{N}_4$ have the same maximal order equal to 5 and use $O(N^3)$ and $O(N^5)$ scalar function evaluations, respectively.

In Section 4 we show that the iteration $I_{-1,1}$ has a smaller cost than any iteration $I_{-1,s}$, $s \ge 2$ and any interpolatory iteration which uses the standard information $\mathfrak{N}_k$, $k = 1,2,3,\ldots$.

2. AN INTERPOLATORY-INTEGRAL ITERATION $I_{-1,s}$

Let $\mathfrak{N}_{-1,s}$ be the integral information defined by (1.2). We want to find an iteration which uses $\mathfrak{N}_{-1,s}$ and has order of convergence as high as possible. Such iterations are called maximal.

Woźniakowski [75] has proved that the maximal order is equal to the order of information. So we wish to find y_d such that the order of information is maximized. Let us briefly recall the ideas of the order of information and the order of iteration. Let $\mathfrak{F}$ be a class of functions F,

$$F: D_F \to B_2,\ D_F \subset B_1,\ \dim(B_1) = \dim(B_2) = N$$

which have a simple zero $\alpha = \alpha(F)$ and are analytic in its neighborhood. Let $\{x_d\}$ be a sequence converging to α, $\lim_d x_d = \alpha$. We shall say that $\{F_d\}$ is equal to $F \in \mathfrak{F}$ with respect to $\mathfrak{N}_{-1,s}$ iff

(2a) $\{F_d\} \subset \mathfrak{F},\ F_d(\alpha_d) = 0,\ \lim_d \alpha_d = \alpha,$

(2b) $\lim_d F_d^{(k)}(\alpha) = G^{(k)}(\alpha),\quad k = 0,1,\ldots,$
where $G \in \mathfrak{F},\ G(\alpha) = 0,$

(2c) $\mathfrak{N}_{-1,s}(x_d;F) = \mathfrak{N}_{-1,s}(x_d;F)\quad \forall d$, i.e.,
$F^{(k)}(x_d) = F_d^{(k)}(x_d),\quad k = 0,1,\ldots,s,$
$\int_0^1 F(x_d + ty_d)dt = \int_0^1 F_d(x_d + ty_d)dt$

The order of information $p = p(\mathfrak{N}_{-1,s})$ is a real number such that

$$p(\mathfrak{N}_{-1,s}) = \begin{cases} \sup A & \text{if } A \neq \emptyset \\ 0 & \text{otherwise} \end{cases}$$

where

$$A = \Big\{p \geq 1: \forall \{x_d\},\ \lim_d x_d = \alpha,\ \forall F \in \mathfrak{F},\ F(\alpha) = 0,\ \forall \{F_d\} \text{ equal to } F \text{ it is true that } \overline{\lim_d} \frac{\|\alpha_d - \alpha\|}{\|x_d - \alpha\|^{p-\varepsilon}} = 0,\ \forall \varepsilon > 0\Big\}.$$

Let $\varphi_{-1,s}$ be an iteration which uses the information $\mathfrak{N}_{-1,s}$. We shall use the notation $h_d = \varphi_{-1,s}(x_d;F)$ which means that h_d is the approximation of α obtained by one step of $\varphi_{-1,s}$ based on x_d and the information $\mathfrak{N}_{-1,s}$.

The order of iteration $\varphi_{-1,s}$, $p = p(\varphi_{-1,s})$ is a real number such that

$$p(\varphi_{-1,s}) = \begin{cases} \sup B & \text{if } B \neq 0 \\ 0 & \text{otherwise} \end{cases}$$

where

$B = \{p \geq 1: \forall \{x_d\}, \lim_d x_d = \alpha, \forall F \in \mathfrak{F}, F(\alpha) = 0, \forall \{F_d\}$ equal F it is true that

$$\overline{\lim_d} \frac{\|h_d - \alpha_d\|}{\|x_d - \alpha\|^{p-\epsilon}} = 0, \ \forall \epsilon > 0$$

where $h_d = \varphi_{-1,s}(x_d;F)\}$

In this section the maximal iteration $I_{-1,s}$ which uses the information $\mathfrak{N}_{-1,s}$ for the solution of (1.1) is defined and the character of convergence is given.

Let x_d be an approximation to a zero α of F. The next approximation x_{d+1} in $I_{-1,s}$ is defined as a zero of the polynomial $u_d = u_d(y;F)$

(2.1) $u_d(x_{d+1};F) = 0$

(with the criterion of its choice, e.g., the nearest zero to x_d), where u_d is defined as follows.

For $N = 1, u_d$ is the unique interpolatory polynomial which agrees with F with respect to the information $\mathfrak{N}_{-1,s}$ for $y_d = \frac{s+3}{s+2}(z_d - x_d)$, $z_d = x_d - \frac{F(x_d)}{F'(x_d)}$. This case is considered in detail by Kacewicz [75], e.g., it is shown there that $I_{-1,s}$ is of order $s + 3$.

For $N \geq 2$ u_d is given by the formula

(2.2) $$u_d(y;F) = F(x_d) + F'(x_d)(y - x_d) + \dots + \frac{1}{s!}F^{(s)}(x_d) \cdot (y - x_d)^s + (s + 2)\left(\frac{s+2}{s+3}\right)^{s+1}\left[\int_0^1 F(x_d + ty_d)dt - F(x_d) - \frac{1}{2} F'(x_d)y_d - \dots - \frac{1}{(s+1)!} F^{(s)}(x_d)y_d^s\right],$$

where

$$(2.3)\quad y_d = \frac{s+3}{s+2}(z_d - x_d), \quad z_d = I_{0,s}(x_d;F),$$

$I_{0,s}$ is the maximal interpolatory iteration which uses the standard information $\mathfrak{N}_s$, thus $\|z_d-\alpha\| = O(\|\alpha-x_d\|^{s+1})$ (see Woźniakowski [74]).

From (2.2) we have

$$(2.4)\quad F(y) - u_d(y;F) = R(y)$$

where

$$(2.5)\quad R(y) = \sum_{k=0}^{2} \frac{1}{(s+1+k)!} F^{(s+1+k)}(x_d)(y - x_d)^{s+1+k} + O(\|y - x_d\|^{s+4}) - (s+2)\left(\frac{s+2}{s+3}\right)^{s+1}\left[\sum_{k=0}^{2} \frac{1}{(s+2+k)!} F^{(s+1+k)}(x_d)y_d^{s+1+k} + O(\|y_d\|^{s+4})\right].$$

From (2.5) and (2.3) we get

$$(2.6)\quad R(y) = \frac{1}{(s+1)!}\left[F^{(s+1)}(x_d)(y - x_d)^{s+1} - F^{(s+1)}(x_d)\cdot(z_d - x_d)^{s+1}\right] + \frac{1}{(s+2)!}\left[F^{(s+2)}(x_d)(y - x_d)^{s+2} - F^{(s+2)}(x_d)(z_d - x_d)^{s+2}\right] + \frac{1}{(s+3)!} F^{(s+3)}(x_d)(y - x_d)^{s+3} - \frac{(s+3)^2}{(s+4)!(s+2)} F^{(s+3)}(x_d)(z_d - x_d)^{s+3} + O(\|y-x_d\|^{s+4}) + O(\|z_d-x_d\|^{s+4}).$$

From (2.1), (2.4), (2.6) and the Schauder fixed point theorem (see Ortega and Rheinboldt [70]) there follows immediately Theorem 1 which gives the character of convergence of the iteration $I_{-1,s}$ for $N \geq 2$.

As part of this theorem we also state the result for $N = 1$ (Kacewicz [75]).

Theorem 1

Let the iteration $I_{-1,s}$ be defined by (2.1). If the function F is sufficiently smooth in the neighborhood of its simple zero α then the sequence $h_d = I_{-1,s}(x_d;F)$ is well defined for x_d sufficiently close to α and for $2 \le N \le +\infty$

$$\overline{\lim_{x_d\to\alpha}} \frac{\|\alpha-h_d\|}{\|\alpha-x_d\|^{s+3-\delta}} \le \begin{cases} \frac{1}{2}\|[F'(\alpha)]^{-1} F''(\alpha)\|^2 & \text{if } s = 1 \\ \delta_{s,2}\left[\frac{1}{(s+1)!}\|[F'(\alpha)]^{-1}F^{(s+1)}(\alpha)\|\right]^2 (s+1) + \\ + \left|1 - \frac{(s+3)^2}{(s+2)(s+4)}\right| \cdot \frac{\|[F'(\alpha)]^{-1}F^{(s+3)}(\alpha)\|}{(s+3)!} & \text{otherwise} \end{cases}$$

where

$$\delta_{i,j} = \begin{cases} 1 & \text{if } i = j \\ 0 & \text{otherwise .} \end{cases}$$

For N = 1

$$\lim_{x_d\to\alpha} \frac{h_d-\alpha}{(x_d-\alpha)^{s+3}} = (-1)^{s+2}\left\{\frac{f''(\alpha)\cdot f^{(s+2)}(\alpha)}{2\cdot(f'(\alpha))^2(s+2)!} + \frac{f^{(s+3)}(\alpha)}{(s+4)!}\cdot\frac{1}{(s+2)f'(\alpha)}\right\}$$

■

Note that in the above theorem $\{x_d\}$ is an arbitrary sequence converging to α, hence we can replace x_d, h_d by the continuous variables x, h, respectively.

3. ORDER OF THE INTEGRAL INFORMATION

In this section we prove that the iteration $I_{-1,s}$ has the order $s + 3 - \delta$ which is as high as possible, i.e., $I_{-1,s}$ is maximal and that y_d given by (2.3) is chosen optimally.

This results follow from the theorem concerning the order of the integral information $p(\mathfrak{N}_{-1,s})$.

Theorem 2

Let $1 \le N \le +\infty$, $s \ge 1$ and

$$\mathfrak{N}_{-1,s} = \mathfrak{N}_{-1,s}(x_d;F) = \left\{F(x_d), F'(x_d), \ldots, F^{(s)}(x_d), \int_0^1 F(x_d+ty_d)dt\right\}$$

where $\quad y_d = y_d(x_d, F(x_d), \ldots, F^{(s)}(x_d))$.

Then

$$p(\mathfrak{N}_{-1,s}) \le s+3-\delta \text{ where } \delta = \begin{cases} 0 & \text{if } N=1 \text{ or } s \ge 2 \\ 1 & \text{otherwise.} \end{cases}$$

Furthermore, if

$$y_d = \frac{s+3}{s+2}(z_d - x_d) \quad \text{for } z_d = I_{0,s}(x_d;F)$$

then

$$p(\mathfrak{N}_{-1,s}) = s+3-\delta.$$

■

Proof

The proof in the case $N = 1$ is omitted since it is given by Kacewicz [75].

Let $N \ge 2$.

We shall prove the first part of Theorem 2. Since the inequality $p(\mathfrak{N}_{-1,s}) \le s+3$ holds for $N = 1$ it also holds for any $2 \le N \le +\infty$. Hence we have to prove that $p(\mathfrak{N}_{-1,1}) \le 3$ and it suffices to consider $N < +\infty$.

Let us define

$$z_d = x_d + \frac{3}{4}y_d, \ \forall d.$$

Case I. $\quad \overline{\lim_d} \frac{\|z_d-\alpha\|}{\|x_d-\alpha\|} > 0$ for certain $F \in \mathfrak{F}$, $F(\alpha) = 0$,

$$\{x_d\}, \ \lim_d x_d = \alpha.$$

Without loss of generality we can assume that $\overline{\lim_d} \frac{\|\alpha-z_d\|}{|\alpha_2-z_{2d}|} > 0$ and $\overline{\lim_d} \frac{\|\alpha-x_d\|}{|\alpha_1-x_{1d}|} > 0$.

Define

$$F_d(x) = F(x) + [(x_1-x_{1d})^2(x_2-z_{2d}),\underbrace{0,\ldots,0}_{N-1}]^T, \quad \forall d$$

where, in general, m_i and m_{id} denote the i-th components of the vectors m, m_d, respectively.

One can verify that $\langle F_d \rangle$ is equal to F with respect to $\mathfrak{N}_{-1,s}$ and

$$(3.1) \quad \overline{\lim_d} \frac{\|\alpha-\alpha_d\|}{\|\alpha-x_d\|^3} > 0,$$

where $F_d(\alpha_d) = 0$.

<u>Case II</u>. $\overline{\lim_d} \dfrac{\|z_d-\alpha\|}{\|x_d-\alpha\|} = 0$ for any F and $\langle x_d \rangle$.

For any $F \in \mathfrak{F}$, $F(\alpha) = 0$ let

$$\text{(i)} \quad \lim_d x_d = \alpha,\ x_{1d} \neq \alpha_1, x_{2d} \neq \alpha_2, \lim_d \frac{\alpha_1-x_{1d}}{\alpha_2-x_{2d}} = 1,\ x_{id} = \alpha_i$$

for $i = 3,4,\ldots,N$.

From here it follows that y_{2d} can be equal 0 only for finite number of d.

Hence without loss of generality we can assume that $y_{2d} \neq 0$, $\forall d$.

We set

$$(3.2) \quad F_d(x) = F(x) + \left[(x_1-x_{1d})^2 - \frac{y_{1d}^2}{y_{2d}^2}(x_2-x_{2d})^2, \underbrace{0,\ldots,0}_{N-1}\right]^T,$$

$$\forall x \in D_F, \ \forall d.$$

One can verify that $\langle F_d \rangle$ is equal F.

Let $a_d = \dfrac{\alpha_1-x_{1d}}{\alpha_2-x_{2d}}$ $(\lim_d a_d = 1)$.

From (3.2) it follows that

$$(3.3)\quad \|\alpha-\alpha_d\| = c_d\|F_d(\alpha)\| = c_d|a_d(z_{2d}-\alpha_2) - (z_{1d}-\alpha_1)|\cdot$$

$$\cdot\ |a_d(z_{2d}-\alpha_2) + (z_{1d}-\alpha_1) + 2(\alpha_1-x_{1d})|,$$

where $\overline{\lim_d}\ c_d = c > 0$, $F_d(\alpha_d) = 0$.

One can verify that there exists a function F and $\{x_d\}$ satisfying the (i) condition such that

$$(3.4)\quad \overline{\lim_d}\ \frac{|a_d(z_{2d}-\alpha_2) - (z_{1d}-\alpha_1)|}{\|\alpha-x_d\|^2} > 0.$$

Indeed, otherwise the iteration φ for the solution of scalar equation $f(r) = 0$ defined as follows

$$\beta_{d+1} = \varphi(\beta_d;f) = z_{2d}(x_d,F(x_d),F'(x_d))-z_{1d}(x_d,F(x_d),F'(x_d))$$

where $F(x) = [x_1,f(x_2),x_3,\ldots,x_N]^T$

and $x_d = \left[\frac{f(\beta_d)}{f'(\beta_d)},\ \beta_d,\ \underbrace{0,\ldots,0}_{N-2}\right]^T$

has order of convergence greater than the order of used information, which is a contradiction.

Hence we proved that for arbitrary $y_d=y_d(x_d,F(x_d),F'(x_d))$ there exist $F \in \mathfrak{J}$, $F(\alpha) = 0$, $\{x_d\}$, $\lim_d x_d = \alpha$, $\{F_d\}$ equal to F such that

$$\overline{\lim_d}\ \frac{\|\alpha-\alpha_d\|}{\|x_d-\alpha\|^3} > 0,$$

which means that $p(\mathfrak{N}_{-1,1}) \leq 3$. This proves the first part of Theorem 2. We shall prove the second part of Theorem 2.

Let $y_d = \frac{s+3}{s+2}(z_d - x_d)$, $z_d = I_{0,s}(x_d;F)$,

where $\lim_d x_d = \alpha$, $F(\alpha) = 0$, $F \in \mathfrak{F}$.

For any $\{F_d\}$ equal to F we have

$$F(y) - F_d(y) = F(y) - u_d(y;F) + u_d(y;F_d) - F_d(y),$$

where u_d is given by (2.2).

From the above, (2.4) and (2.6) for $y = \alpha$ we get

$$||\alpha - \alpha_d|| = O(||\alpha - x_d||^{s+3-\delta}),$$

which completes the proof of Theorem 2. ∎

Let $h_d = I_{-1,s}(x_d;F)$, where $F \in \mathfrak{F}$, $F(\alpha) = 0$. From Theorems 1 and 2 it follows that

$$||h_d - \alpha_d|| = O(||x_d - \alpha||^{s+3-\delta})$$

for any $\{x_d\}$, $\lim_d x_d = \alpha$ and $\{F_d\}$ equal to F, $F_d(\alpha_d) = 0$.

Hence the following corollary holds

Corollary

The iteration $I_{-1,s}$ is of order $s+3-\delta$.

Let $\psi_{-1,s}$ be a class of iterations which use information $\mathfrak{N}_{-1,s}$. The iteration $I_{-1,s}$ has the maximal order in the class $\psi_{-1,s}$, i.e.,

$$p(I_{-1,s}) = \sup_{\varphi_{-1,s} \in \psi_{-1,s}} p(\varphi_{-1,s}), \quad s \geq 1.$$ ∎

4. COMPLEXITY INDEX

The complexity index of an iteration φ of the order p is a measure of the total cost of estimation of the solution α of (1.1). It is defined (see Traub and Woźniakowski [75]) by

$$z(\varphi;F) = \frac{c(\mathfrak{N};F) + c(\varphi)}{\log p}$$

where $\mathfrak{N}$ is the used information,

$c(\mathfrak{N};F)$ is the information cost,

$c(\varphi)$ is the combinatory cost.

For the integral information

$$c(\mathfrak{N}_{-1,s};F) = c(I) + c(\mathfrak{N}_s;F)$$

where $c(I)$ and $c(\mathfrak{N}_s;F)$ are the costs of the computed integral $\int_0^1 F(x_d + ty_d)dt$ and the standard information $\mathfrak{N}_s$, respectively. We want to compare the cost of $I_{-1,s}$ with the cost of the interpolatory iteration $I_{0,k}$ which uses the standard information $\mathfrak{N}_k$ and has order $k + 1$. $I_{-1,s}$ is better than $I_{0,k}$, iff

$$z(I_{-1,s};F) < z(I_{0,k};F), \text{ i.e.,}$$

$$c(I) < \frac{\log(s+3-\delta)}{\log(k+1)} c(\mathfrak{N}_k;F) - c(\mathfrak{N}_s;F) + \frac{\log(s+3-\delta)}{\log(k+1)} c(I_{0,k}) - c(I_{-1,s}). \tag{4.1}$$

Let $c(F^{(i)})$ denote the cost of computing $F^{(i)}(x)$. This cost can be measured by the total number of arithmetical operations needed to compute $F^{(i)}(x)$ as well as by the cost of data access. Note that in most recent computers the cost of data access of indexed variables exceeds the cost of a single arithmetical operation. Let $2 \le N < +\infty$. Let $c(F) = N$. Then it is reasonable to assume that $c(I) = O(N)$. Since $F^{(i)}(x)$ can be represented by $O(N^{i+1})$ scalar function evaluations it seems natural to assume that $c(F^{(i)})$ is comparable with the cost of $O(N^{i+1})$ scalar function evaluations. Thus, let $c(F^{(i)}) = O(N^{i+1})$, $\forall i \ge 1$. It is easy to see that the combinatory costs $c(I_{0,k})$ and $c(I_{-1,s})$ are increasing functions of k and s, respectively.

Then we have

$$\min_{k\ge 1} z(I_{0,k};F) = \min_{k\ge 1} \frac{c(\mathfrak{N}_k;F)+c(I_{0,k})}{\log(k+1)} = z(I_{0,1};F),$$

and

$$\min_{s\ge 1} z(I_{-1,s};F) = \min_{s\ge 1} \frac{c(\mathfrak{N}_s;F)+c(I)+c(I_{-1,s})}{\log(s+3-\delta)} = z(I_{-1,1};F).$$

This implies that the Newton iteration $I_{0,1}$ and the iteration $I_{-1,1}$ are optimal for the problem $F(x) = 0$ in the classes $\{I_{0,k}\}_{k=1,2,\ldots}$ and $\{I_{-1,s}\}_{s=1,2,\ldots}$ respectively. Since $(\log 3-1)c(\mathfrak{N}_1,F) + \log 3\; c(I_{0,1}) - c(I_{-1,1}) = O(N^2)$, (4.1) holds for large N which means that $I_{-1,1}$ is better than the Newton iteration, and hence better than any iteration $I_{0,k}$, $k \ge 1$.

5. INTEGRAL INFORMATION WITH KERNELS

Finally we shall discuss a more general type of integral information, i.e., integral information with kernels

$$\mathfrak{N}_{-1,s}{}^{g} = \left\{F(x_d), F'(x_d), \ldots, F^{(s)}(x_d), \int_0^1 g(t)F(x_d+ty_d)dt\right\}$$

where

$g = g(t)$ is a complex function of a complex variable such

that $\int_0^1 |g(t)|dt < +\infty$, $y_d = y_d(x_d, F(x_d), \ldots, F^{(s)}(x_d))$, $s \geq 1$.

Let $I_j = \int_0^1 g(t)t^{s+j}dt$ and let $m = m(g)$ be the integer defined as follows.

$$m = \begin{cases} 0 \text{ if } I_1 = 0 \\ 1 \text{ if } I_1 \neq 0,\ I_2 = 0 \\ k \text{ if } I_1 \neq 0,\ I_2 \neq 0,\ \dfrac{I_i}{I_1} = \left(\dfrac{I_2}{I_1}\right)^{i-1} \quad \text{for } i = 2,3,\ldots,k \end{cases}$$

$$\text{and } \frac{I_{k+1}}{I_1} \neq \left(\frac{I_2}{I_1}\right)^k,\ k \geq 2.$$

There exists an iteration $I_{-1,s}^{\ g}$ which uses information $\mathfrak{N}_{-1,s}^{\ g}$ for the suitable chosen y_d,

$$y_d = \begin{cases} \dfrac{I_1}{I_2}(z_d - x_d) & \text{if } m \geq 2 \\ z_d - x_d & \text{otherwise, } z_d = I_{0,s}(x_d;F) \end{cases}$$

such that

1. y_d is optimal
2. $p(I_{-1,s}^{\ g}) = \begin{cases} \min(s+1+m, 2s+2) & \text{if } N = 1 \\ \min(s+1+m, 2s+1) & \text{if } 2 \leq N \leq +\infty \end{cases}$
3. $I_{-1,s}^{\ g}$ is maximal
4. There exists $g = g(t)$ such that

$$p(I_{-1,s}^{\ g}) = \begin{cases} 2s+2 & \text{if } N = 1 \\ 2s+1 & \text{if } 2 \leq N \leq +\infty \end{cases}$$

The proof is based on techniques similar to those used here. These results will be reported in a future paper.

ACKNOWLEDGMENT

I wish to express my appreciation to J. F. Traub and H. Woźniakowski for their helpful comments during the preparation of this paper.

6. REFERENCES

Kacewicz [75] Kacewicz, B., "An Integral-Interpolatory Iterative Method for the Solution of Non-Linear Scalar Equations," Computer Science Department Report, Carnegie-Mellon University, Pittsburgh, Pa., 1975.

Ortega and Rheinboldt [70] Ortega, J. M. and W. C. Rheinboldt, Iterative Solution of Nonlinear Equations in Several Variables, Academic Press, New York and London, 1970.

Traub [64] Traub, J. F., Iterative Methods for the Solution of Equations, Prentice Hall, 1964.

Traub and Woźniakowski [75] Traub, J. F. and H. Wozniakowski, "Strict Lower and Upper Bounds on Iterative Complexity," these Proceedings.

Woźniakowski [74] Woźniakowski, H., "Maximal Stationary Iterative Methods for the Solution of Operator Equations," SIAM J. Numer. Anal., Vol. 11, No. 5, October 1974, 934-949.

Woźniakowski [75] Woźniakowski, H., "Generalized Information and Maximal Order of Iteration for Operator Equations," SIAM J. Numer. Anal., Vol. 12, No. 1, March 1975, 121-135.

Complexity and Differential Equations*

M. H. SCHULTZ

Department of Computer Science
Yale University
New Haven, Connecticut

§1. INTRODUCTION

In this paper we briefly survey three analytic complexity topics, which are the subject of active research in the Department of Computer Science of Yale University. Much of the work described has been done jointly with S. C. Eisenstat and A. H. Sherman.

S. Winograd suggested the first topic, dealing with the complexity of interpolation problems. In §2, we extend the class of problems considered in Rivlin and Winograd [74] and Winograd [75] and obtain both lower and upper bounds for this new class of "generalized interpolation problems."

In §3, we discuss the "optimal storage algorithms," introduced in Eisenstat, Schultz, and Sherman [75b], for the direct solution of sparse linear systems of equations. We pay particular attention to the application of these algorithms to the 5-point difference equations that arise as approximations to partial differential equations in two independent variables. This class of algorithms is a good example of the general phenomenon of being able to construct algorithms for particular problems that trade off computing time for storage. In this case, we can gain an order of magnitude improvement in the asymptotic storage requirement without affecting the asymptotic time requirement.

Our third topic, discussed in §4, deals with a special case of the following general problem for finding fixed points. Suppose we wish to compute $\underline{x}$ such that $\underline{x} = \underline{\phi}(\underline{x})$. Find functions $\underline{\phi}_k$, k = 1,2,... such that the iterative method

---

* This work was supported in part by the National Science Foundation under Grant GJ-43157 and the Office of Naval Research under Grant N0014-67-A-0097-0016.

(1) $\underset{\sim}{x}_1$ arbitrary

(2) $\underset{\sim}{x}_{k+1} = \Phi_k(\underset{\sim}{x}_k)$, $k = 1,2,\ldots$

generates a sequence of iterates $\underset{\sim}{x}_k$ that converge "as fast as possible." Historically, "as fast as possible" has meant in terms of the number of iterations or function evaluations required to obtain an approximation with a prescribed accuracy.

However, the problem is actually much more subtle in that we can, in addition to choosing the Φ_k's, vary the precision with which we compute the iterates $\underset{\sim}{x}_{k+1}$ (or the iteration function Φ_k) in order to maintain the convergence properties of the original iteration scheme but to minimize the work. The particular example we will consider is the application of Newton's method to solve the system of sparse nonlinear equations arising from the application of a finite difference method to approximate the solution of partial differential equations. We approximate $\underset{\sim}{x}_{k+1}$ at each step by means of an inner-iterative procedure with fixed stopping criteria.

Under these special circumstances we give some surprising work estimates, which first appeared in Eisenstat, Schultz, and Sherman [74a], and whose complete derivation is given in Sherman [75]. These estimates indicate that the study of the complexity of the "general iteration problem" for nonlinear systems may yield surprising results, which will have a significant effect on practical computing.

§2. GENERALIZED INTERPOLATION PROBLEMS

The discussion in this section is a generalization of the ideas presented in Rivlin and Winograd [74] and Winograd [75]. Further details and the proofs will appear in Schultz [75].

Let $X \subset Y$ be two Banach spaces and T be a bounded linear mapping of X into Y. Given a finite set Λ of linearly independent, continuous linear functionals

$$\Lambda \equiv \{\lambda_i \mid 1 \le i \le k\} \subset X^* \tag{2.1}$$

on X and a k-vector $\underline{r} \in E^k$, the *generalized interpolation problem* is to find an element $x \in X$ such that

$$\lambda_i(x) = r_i, \text{ for all } 1 \le i \le k. \tag{2.2}$$

We let $S(\underline{r})$ denote the set of all solutions of the generalized interpolation problem, i.e.

$$S(\underline{r}) \equiv \{x \in X \mid \lambda_i(x) = r_i,\ 1 \le i \le k\}. \tag{2.3}$$

Moreover, we assume that $TS(\underline{0})$ is closed in Y. Conditions under which this is true are discussed in Schultz [75].

In order to study the computational complexity of the generalized interpolation problem, we define the idea of Λ-*recovery scheme* as a mapping $R_\Lambda: E^k \to X$ such that

$$(2.4) \qquad R_\Lambda(\underline{r}) \in S(\underline{r}), \text{ for all } \underline{r} \in E^k.$$

The fundamental question is what can we say about

$$(2.5) \qquad E(\Lambda) \equiv \inf_{R_\Lambda} \sup_{\substack{f \in X \\ \|Tf\|_y \le 1}} \|R_\Lambda(\Lambda(f)) - f\|_y ,$$

where $\Lambda(f)$ is the k-vector whose *ith* component is $\lambda_i(f)$.

The answer is given by the following approximation theory type result from Schultz [75].

Theorem 2.1. If $h \equiv \inf_{\substack{g \in S(\underline{0}) \\ g \neq 0}} \|Tg\|_y / \|g\|_y$

and $c \equiv \begin{cases} h^{-1} & \text{if } h \neq 0 \\ \infty & \text{if } h = 0 \end{cases}$, then

$$(2.6) \qquad c \le E(\Lambda) \le 2c.$$

Moreover, if Y is a Hilbert space

$$(2.7) \qquad E(\Lambda) = c.$$

If Y is a Hilbert space, it is shown in Schultz [75] that the optimal recovery scheme $\tilde{R}_\lambda$ is such that $\tilde{R}_\Lambda(r)$ minimizes $\|Tz\|_y$ over all $z \in S(\underline{r})$. Geometrically $\tilde{R}_\Lambda(\underline{r})$ is the best approximation in Y to $\overline{0}$ from the affine space $\{Tz \mid z \in U(r)\}$.

As a prime example, we consider the choices of $X \equiv W^{n,2}[0,1]$, $Y \equiv L^2[0,1]$, $T \equiv D^n$, and $\lambda_i(f) \equiv f(x_i)$, $1 \le i \le k$, where $0 = x_1 < x_2 < \ldots < x_{k-1} < x_k = 1$. Then $\tilde{R}_\Lambda(f)$ can be shown to be the natural spline interpolate of order 2n, cf. Schultz [73].

§3. SPARSE LINEAR SYSTEMS

In this section we consider the problem of solving the linear system

$$(3.1) \qquad A\underline{x} = \underline{k} ,$$

where A is an $n^2 \times n^2$ symmetric, positive-definite band matrix with bandwidth n and at most five nonzero entries on any row of the following form:

(3.2) $A \equiv$  .

The graph associated with the matrix A is the $n \times n$ square grid. For simplicity, we assume that n is of the form 2^t-1, where t is a positive integer.

If we solve (3.1) using symmetric band elimination we need n^3 storage locations and $1/2\ n^4$ multiplications, cf. Martin and Wilkinson [65]. In this algorithm we compute a triangular factorization of the matrix A and solve for the unknowns corresponding to consecutive grid points along consecutive rows.

However, as is common with many other problems, it is better to use a "divide and conquor" strategy. Very briefly, this is accomplished in this problem by first solving for the unknowns associated with the grid points along the middle row by eliminating the unknowns associated with the grid points above and below this row, without storing the factorization. Once we have solved for these unknowns, we have split the problem in half to form two $(2^t-1) \times 2(2^{t-1}-1)$ subproblems. We then divide each of these subproblems by solving for the unknowns along the middle columns, leaving us with four $2(2^{t-1}-1) \times 2(2^{t-1}-1)$ subproblems. Each of these subproblems is identical in structure to the original problem and is subdivided in the same way. We continue until we reach subproblems that are dense systems and are solved by Gaussian elimination. We call this procedure the minimal storage band elimination algorithm. Leaving the details to Sherman [75] and Eisenstat, Schultz, and Sherman [74b] we report only the final outcome.

Theorem 3.1. The minimal storage band elimination algorithm requires n^2 storage locations and $\sim 5/6\ n^4$ multiplications.

Moreover, we can show that using similar techniques in conjunction with sparse elimination we can achieve even better results. In particular, we can obtain an algorithm that requires $O(n^2)$ storage locations and $O(n^3)$ multiplications, cf. Sherman [75] and Eisenstat, Schultz, and Sherman [74b].

It is important to note the existence of the classic lower bound that all factorizations of the form $LL^T = PAP^T$, where P is an $n^2 \times n^2$ permutation matrix, require at least $O(n^2 \log n)$ storage locations, cf. Hoffman, Martin, and Rose [73]. Thus by enlarging our class of admissable algorithms we have been able to break through this classic lower bound.

§4. SPARSE NONLINEAR SYSTEMS

Suppose we wish to solve a nonlinear system

$$(4.1) \qquad \underline{f}(\underline{x}) = \underline{0} ,$$

where the Jacobian matrix $J\underline{f}$ is a symmetric, positive-definite five diagonal matrix such as described in §3. A favorite method for approximating a root of (4.1) is the quadratically convergent Newton's method

$$(4.2) \qquad \underline{x}^{m+1} = \underline{x}^m - [J\underline{f}(\underline{x}^m)]^{-1}f(\underline{x}^m), \ m = 1,2,\ldots ,$$

where $\underline{x}^1$ is arbitrary. If $\underline{x}^1$ is sufficiently close to the root $\underline{x}$, then this method requires log log n iterations to reduce the initial error by a factor of $O(n^{-2})$, which is the usual order of the discretization error when (4.1) is a finite difference approximation to a partial differential equation.

In order actually to carry out Newton's method, typically one factors $J\underline{f}(\underline{x}^m)$ at each iteration and solves for the correction $\underline{x}^{m+1}-\underline{x}^m$. This requires at least $O(n^3)$ multiplications, yielding a total of $O(n^3 \log\log n)$ multiplications to obtain an $O(n^{-2})$ accurate solution.

However, why should we compute $\underline{x}^{m+1}$ exactly, when it itself is only an approximation to $\underline{x}$? Indeed in the scalar case, Brent has asked this question and studied its answer with respect to high precision root finding and applications to high precision function evaluation, cf. Brent [75]. Furthermore, the typical approach outlined above takes no advantage of the fact that $\underline{x}^m$ converges to $\underline{x}$ and hence itself is a good starting guess for an iterative method for computing $\underline{x}^{m+1}$.

In fact, we do better by approximating $\underline{x}^{m+1}$ by an "inner iterative" method. Following Eisenstat, Schultz, and Sherman [74a], we define the "Newton-Sparse-Richardson methods."

Let $\underline{x}^1$ be arbitrary,

$$(4.3) \qquad \underline{z}_k^m \equiv \underline{x}_k^{m+1} - \underline{x}_k^m, \ m = 1,2,\ldots, \ k = 1,2,\ldots ,$$

where $\underline{z}_0^m \equiv 0$,

$$(4.4) \qquad J\underline{f}(\underline{x}^1)\underline{z}_{k+1}^m = J\underline{f}(\underline{x}^1)\underline{z}_k^m + \gamma[J\underline{f}(\underline{x}^m)\underline{z}_k^m+\underline{f}(\underline{x}^m)]$$

and

$$(4.5) \qquad \underline{x}^{m+1} = \underline{x}^m + \underline{z}_{\theta(m)}^m$$

where γ is an appropriately chosen relaxation parameter.

The choice of $\theta(m) = 2^m$ yields a quadratically convergent method, while the choice of $\theta(m) = 1$, which is equivalent to the relaxed chord method, cf. Blum [72], yields a linearly convergent method. Moreover, in either case the matrix $J\underline{f}(\underline{x}^1)$ is fixed throughout the algorithm. Hence, we factor it (using sparse techniques) during the first iteration and thereafter use its factored form.

Superficially one would expect the quadratically convergent variation to be more efficient than the linearly convergent variation. Surprisingly enough, as was shown in Eisenstat, Schultz, and Sherman [74a] and Sherman [75], this is not the case. In fact, both variations require $O(n^3) + O(n^2 (\log n)^2)$ multiplications to reduce the initial error by a factor of $O(n^{-2})$, where the first term is the work to factor $J\underline{f}(\underline{x}^1)$. In practice, the choice of the particular variation to be used depends on the relative cost of evaluations of $J\underline{f}(\underline{y})$ versus evaluations of $\underline{f}(\underline{y})$. Computational results are reported in Eisenstat, Schultz, and Sherman [74a] and Sherman [75].

REFERENCES

Blum [72] E. K. Blum. *Numerical Analysis and Computation: Theory and Practice*. Addison-Wesley, 1972.

Brent [75] R. P. Brent. Multiple-precision zero-finding methods and the complexity of elementary function evaluation. These Proceedings.

Eisenstat, Schultz, and Sherman [74a] S. C. Eisenstat, M. H. Schultz, and A. H. Sherman. The application of sparse matrix methods to the numerical solution of nonlinear elliptic partial differential equations. In A. Dold and B. Eckmann, editors, *Proceedings of the Symposium on Constructive and Computational Methods for Differential Equations*. Springer-Verlag, 131-153, 1974.

Eisenstat, Schultz, and Sherman [74b] S. C. Eisenstat, M. H. Schultz, and A. H. Sherman. Direct methods for the solution of large sparse systems of linear equations with limited core storage. Presented at the SIAM Fall Meeting, Alexandria, Virginia, 1974.

Hoffman, Martin, and Rose [73] A. J. Hoffman, M. S. Martin, and D. J. Rose. Complexity bounds for regular finite difference and finite element grids. *SIAM Journal on Numerical Analysis* 10:364-369, 1973.

Martin and Wilkinson [65] R. S. Martin and J. H. Wilkinson. Symmetric decomposition of positive definite band

matrices. *Numerische Mathematik* 7:355-361, 1965.
Rivlin and Winograd [74] T. J. Rivlin and S. Winograd. The optimal recovery of smooth functions. IBM Research Report RC 4920 (#21869), 1974.
Schultz [73] M. H. Schultz. *Spline Analysis*. Prentice-Hall, 1973.
Schultz [75] M. H. Schultz. Optimal recovery schemes for generalized interpolation problems. To appear.
Sherman [75] A. H. Sherman. On the efficient solution of sparse systems of linear and nonlinear equations. PhD dissertation, Yale University, 1975.
Winograd [75] Optimal approximation from discrete samples. These Proceedings.

MULTIPLE-PRECISION ZERO-FINDING METHODS AND THE COMPLEXITY OF ELEMENTARY FUNCTION EVALUATION

Richard P. Brent

Computer Centre,
Australian National University,
Canberra, A.C.T. 2600, Australia

ABSTRACT

We consider methods for finding high-precision approximations to simple zeros of smooth functions. As an application, we give fast methods for evaluating the elementary functions log(x), exp(x), sin(x) etc. to high precision. For example, if x is a positive floating-point number with an n-bit fraction, then (under rather weak assumptions) an n-bit approximation to log(x) or exp(x) may be computed in time asymptotically equal to $13M(n)\log_2 n$ as $n \to \infty$, where M(n) is the time required to multiply floating-point numbers with n-bit fractions. Similar results are given for the other elementary functions, and some analogies with operations on formal power series are mentioned.

1. INTRODUCTION

When comparing methods for solving nonlinear equations or evaluating functions, it is customary to assume that the basic arithmetic operations (addition, multiplication, etc.) are performed with some fixed precision. However, an irrational number can only be approximated to arbitrary accuracy if the precision is allowed to increase indefinitely. Thus, we shall consider iterative processes using variable precision. Usually the precision will increase as the computation proceeds, and the final result will be obtained to high precision. Of course, we could use the same (high) precision throughout, but then the computation would take longer than with variable precision, and the final result would be no more accurate.

Assumptions

For simplicity we assume that a standard multiple-precision floating-point number representation is used, with a binary fraction of n bits, where n is large. The exponent length is fixed, or may grow as $o(n)$ if necessary. To avoid table-lookup methods, we assume a machine with a finite random-access memory and a fixed number of sequential tape units. Formally, the results hold for multitape Turing machines.

Precision n Operations

An operation is performed with precision n if the operands and result are floating-point numbers as above (i.e., precision n numbers), and the relative error in the result is $O(2^{-n})$.

Precision n Multiplication

Let $M(n)$ be the time required to performe precision n multiplication. (Time may be regarded as the number of single-precision operations, or the number of bit operations, if desired.) The classical method gives $M(n) = O(n^2)$, but methods which are faster for large n are known. Asymptotically the fastest method known is that of Schönhage and Strassen [71], which gives

(1.1) $$M(n) = O(n.\log(n)\log\log(n)) \quad \text{as} \quad n \to \infty.$$

Our results do not depend on the algorithm used for multiplication, provided $M(n)$ satisfies the following two conditions.

(1.2) $n = o(M(n))$, i.e., $\lim_{n\to\infty} n/M(n) = 0$;

and, for any $\alpha > 0$,

(1.3) $M(\alpha n) \sim \alpha M(n)$, i.e., $\lim_{n\to\infty} \frac{M(\alpha n)}{\alpha M(n)} = 1$.

Condition (1.2) enables us to neglect additions, since the time for an addition is $O(n)$, which is asymptotically negligible compared to the time for a multiplication. Condition (1.3) certainly holds if

$$M(n) \sim cn[\log(n)]^{\beta}[\log\log(n)]^{\gamma} ,$$

though it does not hold for some implementations of the Schönhage-Strassen method. We need (1.3) to estimate the constants in the asymptotic "O" results: if the constants are not required then much weaker assumptions suffice, as in Brent [75a,b].

The following lemma follows easily from (1.3).

Lemma 1.1

If $0 < \alpha < 1$, $M(n) = 0$ for $n < 1$, and $c_1 < \frac{1}{1-\alpha} < c_2$, then

$$c_1 M(n) < \sum_{k=0}^{\infty} M(\alpha^k n) < c_2 M(n)$$

for all sufficiently large n .

2. BASIC MULTIPLE-PRECISION OPERATIONS

In this section we summarize some results on the time required to perform the multiple-precision operations of division, extraction of square roots, etc. Additional results are given in Brent [75a].

Reciprocals

Suppose $a \neq 0$ is given and we want to evaluate a precision n approximation to $1/a$. Applying Newton's method to the equation

$$f(x) \equiv a - 1/x = 0$$

gives the well-known iteration

$$x_{i+1} = x_i - x_i \varepsilon_i ,$$

where

$$\varepsilon_i = ax_i - 1 .$$

Since the order of convergence is two, only $k \sim \log_2 n$ iterations are required if x_0 is a reasonable approximation to $1/a$, e.g., a single-precision approximation.

If $\varepsilon_k = O(2^{-n})$, then $\varepsilon_{k-1} = O(2^{-n/2})$, so at the last iteration it is sufficient to perform the multiplication of x_{k-1} by ε_{k-1} using precision $n/2$, even though ax_{k-1} must be evaluated with precision n . Thus, the time required for the last iteration is $M(n) + M(n/2) + O(n)$. The time for the next to last iteration is $M(n/2) + M(n/4) + O(n/2)$, since this iteration need only give an approximation accurate to $O(2^{-n/2})$, and so on. Thus, using Lemma 1.1, the total time required is

$$I(n) \sim (1 + \tfrac{1}{2})(1 + \tfrac{1}{2} + \tfrac{1}{4} + \ldots)M(n) \sim 3M(n)$$

as $n \to \infty$.

Division

Since $b/a = b(1/a)$, precision n division may be done in time

$$D(n) \sim 4M(n)$$

as $n \to \infty$.

Inverse Square Roots

Asymptotically the fastest known method for evaluating $a^{-\frac{1}{2}}$ to precision n is to use the third-order iteration

$$x_{i+1} = x_i - \tfrac{1}{2}x_i(\varepsilon_i - \tfrac{3}{4}\varepsilon_i^2) ,$$

where

$$\varepsilon_i = ax_i^2 - 1 .$$

At the last iteration it is sufficient to evaluate ax_i^2 to precision n , ε_i^2 to precision n/3 , and $x_i(\varepsilon_i - \frac{3}{4}\varepsilon_i^2)$ to precision 2n/3 . Thus, using Lemma 1.1 as above, the total time required is

$$Q(n) \sim (2 + \frac{1}{3} + \frac{2}{3})(1 + \frac{1}{3} + \frac{1}{9} + \ldots)M(n) \sim 4\tfrac{1}{2}M(n)$$

as $n \to \infty$.

<u>Square Roots</u>

Since

$$a^{\frac{1}{2}} = \begin{cases} a.a^{-\frac{1}{2}} & \text{if } a > 0 , \\ 0 & \text{if } a = 0 , \end{cases}$$

we can evaluate $a^{\frac{1}{2}}$ to precision n in time

$$R(n) \sim 5\tfrac{1}{2}M(n)$$

as $n \to \infty$. Note that the direct use of Newton's method in the form

$$x_{i+1} = \frac{1}{2}(x_i + a/x_i) \tag{2.1}$$

or

$$x_{i+1} = x_i + \left(\frac{a - x_i^2}{2x_i}\right) \tag{2.2}$$

is asymptotically slower, requiring time $\sim 8M(n)$ or $\sim 6M(n)$ respectively.

3. VARIABLE-PRECISION ZERO-FINDING METHODS

Suppose $\zeta \neq 0$ is a simple zero of the nonlinear equation

$$f(x) = 0 .$$

Here, $f(x)$ is a sufficiently smooth function which can be evaluated near ζ , with absolute error $0(2^{-n})$, in time $w(n)$. We consider some methods for evaluating ζ to precision n .

Since we are interested in results for very large n, the time required to obtain a good starting approximation is neglected.

Assumptions

To obtain sharp results we need the following two assumptions, which are similar to (1.2) and (1.3):

(3.1) $M(n) = o(w(n))$, i.e., $\lim_{n\to\infty} M(n)/w(n) = 0$;

and, for some $\alpha \geq 1$ and all $\beta > 0$,

(3.2) $$w(\beta n) \sim \beta^{\alpha} w(n)$$

as $n \to \infty$.

From (3.1), the time required for a multiplication is negligible compared to the time for a function evaluation, if n is sufficiently large. (3.2) implies (3.1) if $\alpha > 1$, and (3.2) certainly holds if, for example,

$$w(n) \sim cn^{\alpha}[\log(n)]^{\gamma}[\log\log(n)]^{\delta} .$$

The next lemma follows from our assumptions in much the same way as Lemma 1.1.

Lemma 3.1

If $0 < \beta < 1$, $w(n) = 0$ for $n < 1$, and

$$c_1 < 1/(1 - \beta^{\alpha}) < c_2 ,$$

then

$$c_1 w(n) < \sum_{k=0}^{\infty} w(\beta^k n) < c_2 w(n)$$

for all sufficiently large n.

A Discrete Newton Method

To illustrate the ideas of variable-precision zero-finding methods, we describe a simple discrete Newton method. More efficient methods are described in the next three sections, and in Brent [75a].

Consider the iteration

$$x_{i+1} = x_i - f(x_i)/g_i ,$$

where g_i is a one-sided difference approximation to $f'(x_i)$, i.e.,

$$g_i = \frac{f(x_i + h_i) - f(x_i)}{h_i} .$$

If $\varepsilon_i = |x_i - \zeta|$ is sufficiently small, $f(x_i)$ is evaluated with absolute error $O(\varepsilon_i^2)$, and h_i is small enough that

$$g_i = f'(x_i) + O(\varepsilon_i) , \tag{3.3}$$

then the iteration converges to ζ with order at least two. To ensure (3.3), take h_i of order ε_i , e.g. $h_i = f(x_i)$.

To obtain ζ to precision n, we need two evaluations of f with absolute error $O(2^{-n})$, preceded by two evaluations with error $O(2^{-n/2})$, etc. Thus, the time required is

$$t(n) \sim 2(1 + 2^{-\alpha} + 2^{-2\alpha} + \ldots)w(n) . \tag{3.4}$$

<u>Asymptotic Constants</u>

We say that a zero-finding method has <u>asymptotic constant</u> $C(\alpha)$ if, to find a simple zero $\zeta \neq 0$ to precision n, the method requires time $t(n) \sim C(\alpha)w(n)$ as $n \to \infty$. (The asymptotic constant as defined here should not be confused with the "asymptotic error constant" as usually defined for single-precision zero-finding methods.)

For example, from (3.4), the discrete Newton method described above has asymptotic constant

$$C_N(\alpha) = 2/(1 - 2^{-\alpha}) \lesssim 4 .$$

Note that the time required to evaluate ζ to precision n is only a small multiple of the time required to evaluate $f(x)$

with absolute error $O(2^{-n})$. (If we used fixed precision, the time to evaluate ζ would be $O(\log(n))$ times the time to evaluate $f(x)$.)

4. A VARIABLE-PRECISION SECANT METHOD

The secant method is known to be more efficient than the discrete Newton method when fixed-precision arithmetic is used. The same is true with variable-precision arithmetic, although the ratio of efficiencies is no longer constant, but depends on the exponent α in (3.2). Several secant-like methods are described in Brent [75a]; here we consider the simplest such method, which is also the most efficient if $\alpha < 4.5243\ldots$.

The secant iteration is

$$x_{i+1} = x_i - f_i \left(\frac{x_i - x_{i-1}}{f_i - f_{i-1}}\right) ,$$

where $f_i = f(x_i)$, and we assume that the function evaluations are performed with sufficient accuracy to ensure that the order of convergence is at least $\rho = \frac{1}{2}(1 + 5^{\frac{1}{2}}) = 1.6180\ldots$, the larger root of

$$\rho^2 = \rho + 1 . \tag{4.1}$$

Let $\varepsilon = |x_{i-1} - \zeta|$. Since the smaller root of (4.1) lies inside the unit circle, we have

$$x_i - \zeta = O(\varepsilon^{\rho})$$

and

$$x_{i+1} - \zeta = O(\varepsilon^{\rho^2}) .$$

To give order ρ , f_i must be evaluated with absolute error $O(\varepsilon^{\rho^2})$. Since $f_i = O(|x_i - \zeta|) = O(\varepsilon^{\rho})$, it is also necessary to evaluate $(f_i - f_{i-1})/(x_i - x_{i-1})$ with relative error $O(\varepsilon^{\rho^2 - \rho})$, but $|x_i - x_{i-1}| \sim \varepsilon$, so it is necessary to evaluate f_{i-1} with absolute error $O(\varepsilon^{\rho^2 - \rho + 1})$. [Since

f_i must be evaluated with absolute error $O(\varepsilon^{\rho^2})$, f_{i-1} must be evaluated with absolute error $O(\varepsilon^{\rho})$, but $\rho^2 - \rho + 1 = 2 > \rho$, so this condition is superfluous.]

The conditions mentioned are sufficient to ensure that the order of convergence is at least ρ . Thus, if we replace ε^{ρ^2} by 2^{-n} , we see that ζ may be evaluated to precision n if f is evaluated with absolute errors $O(2^{-n})$, $O(2^{-2n\rho^{-2}})$, $O(2^{-2n\rho^{-3}})$, $O(2^{-2n\rho^{-4}})$, ... It follows that the asymptotic constant of the secant method is

$$C_S(\alpha) = 1 + (2\rho^{-2})^{\alpha}/(1 - \rho^{-\alpha}) \leqslant C_S(1) = 3 .$$

The following lemma states that the secant method is asymptotically more efficient than the discrete Newton method when variable precision is used.

<u>Lemma 4.1</u>

$C_S(\alpha) < C_N(\alpha)$ for all $\alpha \geqslant 1$. In fact, $C_S(\alpha)/C_N(\alpha)$ decreases monotonically from $\frac{3}{4}$ (when $\alpha = 1$) to $\frac{1}{2}$ (as $\alpha \to \infty$).

5. OTHER VARIABLE-PRECISION INTERPOLATORY METHODS

With fixed precision, inverse quadratic interpolation is more efficient than linear interpolation, and inverse cubic interpolation is even more efficient, if the combinatory cost (i.e., "overhead") is negligible. With variable precision the situation is different. Inverse quadratic interpolation is slightly more efficient than the secant method, but inverse cubic interpolation is <u>not</u> more efficient than inverse quadratic interpolation if $\alpha \lesssim 4.6056...$. Since the combinatory cost of inverse cubic interpolation is considerably higher than that of inverse quadratic interpolation, the inverse cubic method appears even worse if combinatory costs are significant.

Inverse Quadratic Interpolation

The analysis of variable-precision methods using inverse quadratic interpolation is similar to that for the secant method, so we only state the results. The order $\rho = 1.8392\ldots$ is the positive root of $\rho^3 = \rho^2 + \rho + 1$. It is convenient to define $\sigma = 1/\rho = 0.5436\ldots$ To evaluate ζ to precision n requires evaluations of f to (absolute) precision n, $(1 - \sigma + \sigma^2)n$, and $\sigma^j(1 - \sigma - \sigma^2 + 2\sigma^3)n$ for $j=0,1,2,\ldots$ Thus, the asymptotic constant is

$$C_Q(\alpha) = 1 + (1 - \sigma + \sigma^2)^\alpha + (3\sigma^3)^\alpha/(1 - \sigma^\alpha)$$
$$\leqslant C_Q(1) = \tfrac{1}{2}(7 - 2\sigma - \sigma^2) = 2.8085\ldots .$$

Lemma 5.1

$C_Q(\alpha) < C_S(\alpha)$ for all $\alpha \geqslant 1$. In fact, $C_Q(\alpha)/C_S(\alpha)$ increases monotonically from $0.9361\ldots$ (when $\alpha = 1$) to 1 (as $\alpha \to \infty$).

Inverse Cubic Interpolation, etc.

If $\mu = 0.5187\ldots$ is the positive root of $\mu^4 + \mu^3 + \mu^2 + \mu = 1$, then the variable-precision method of order $1/\mu = 1.9275\ldots$, using inverse cubic interpolation, has asymptotic constant

$$C_C(\alpha) = 1 + (1 - \mu + \mu^2)^\alpha + (1 - \mu - \mu^2 + 2\mu^3)^\alpha$$
$$+ (4\mu^4)^\alpha/(1 - \mu^\alpha)$$
$$\leqslant C_C(1) = (13 - 6\mu - 4\mu^2 - 2\mu^3)/3 = 2.8438\ldots .$$

Note that $C_C(1) > C_Q(1)$. Variable-precision methods using inverse interpolation of arbitrary degree are described in Brent [75a]. Some of these methods are slightly more efficient than the inverse quadratic interpolation method if α is large, but inverse quadratic interpolation is the most efficient method known for $\alpha < 4.6056\ldots$. In practice α

is usually 1, $1\frac{1}{2}$ or 2.

An Open Question

Is there a method with asymptotic constant $C(\alpha)$ such that $C(1) < C_Q(1)$?

6. VARIABLE-PRECISION METHODS USING DERIVATIVES

In Sections 3 to 5 we considered methods for solving the nonlinear equation $f(x) = 0$, using only evaluations of f . Sometimes it is easy to evaluate $f'(x)$, $f''(x)$, ... once $f(x)$ has been evaluated, and the following theorem shows that it is possible to take advantage of this. For an application, see Section 10.

Theorem 6.1

If the time to evaluate $f(x)$ with an absolute error $O(2^{-n})$ is $w(n)$, where $w(n)$ satisfies conditions (3.1) and (3.2), and (for $k=1,2,\ldots$) the time to evaluate $f^{(k)}(x)$ with absolute error $O(2^{-n})$ is $w_k(n)$, where

$$w_k(n) = o(w(n))$$

as $n \to \infty$, then the time to evaluate a simple zero $\zeta \neq 0$ of $f(x)$ to precision n is

$$t(n) \sim w(n)$$

as $n \to \infty$.

Proof

For fixed $k \geq 1$, we may use a direct or inverse Taylor series method of order $k + 1$. The combinatory cost is of order $k.\log(k + 1).M(n)$ (see Brent and Kung [75]). From (3.1), this is $o(w(n))$ as $n \to \infty$. Thus,

$$\begin{aligned} t(n) &\leq [1 - (k + 1)^{-\alpha}]^{-1} w(n) + o(w(n)) \\ &\leq (1 + \frac{1}{k} + o(1))w(n) . \end{aligned}$$

For sufficiently large n, the "$o(1)$" term is less than $1/k$, so

$$t(n) \lesssim (1 + \frac{2}{k})w(n) .$$

Given $\varepsilon > 0$, choose $k \gtrsim 2/\varepsilon$. Then, for all sufficiently large n,

$$w(n) \lesssim t(n) \lesssim (1 + \varepsilon)w(n) ,$$

so $t(n) \sim w(n)$ as $n \to \infty$.

Corollary 6.1

If the conditions of Theorem 6.1 hold, $f:[a,b] \to I$, $f'(x) \neq 0$ for $x \in [a,b]$, and g is the inverse function of f, then the time to evaluate $g(y)$ with absolute error $O(2^{-n})$, for $y \in I$, is

$$w_g(n) \sim w(n)$$

as $n \to \infty$.

Note

Corollary 6.1 is optimal in the sense that, if $w_g(n) \sim cw(n)$ for some constant $c < 1$, then $w(n) \sim cw_g(n)$ by the same argument, so $w(n) \sim c^2 w(n)$, a contradiction. Hence, $c = 1$ is minimal.

7. THE ARITHMETIC-GEOMETRIC MEAN ITERATION

Before considering the multiple-precision evaluation of elementary functions, we recall some properties of the arithmetic-geometric (A-G) mean iteration of Gauss [1876]. Starting from any two positive numbers a_0 and b_0, we may iterate as follows:

$$a_{i+1} = \frac{1}{2}(a_i + b_i) \qquad \text{(arithmetic mean)}$$

and

$$b_{i+1} = (a_i b_i)^{\frac{1}{2}} \qquad \text{(geometric mean)}$$

for $i=0,1,\ldots$.

Second-order Convergence

The A-G mean iteration is of computational interest because it converges very fast. If $b_i \ll a_i$, then

$$b_{i+1}/a_{i+1} = \frac{2(b_i/a_i)^{\frac{1}{2}}}{1 + b_i/a_i} \simeq 2(b_i/a_i)^{\frac{1}{2}} ,$$

so only about $|\log_2(a_0/b_0)|$ iterations are required before a_i/b_i is of order 1 . Once a_i and b_i are close together the convergence is second order, for if $b_i/a_i = 1 - \varepsilon_i$ then

$$\varepsilon_{i+1} = 1 - b_{i+1}/a_{i+1} = 1 - 2(1 - \varepsilon_i)^{\frac{1}{2}}/(2 - \varepsilon_i) = \varepsilon_i^2/8 + O(\varepsilon_i^3) .$$

Limit of the A-G Mean Iteration

There is no essential loss of generality in assuming that $a_0 = 1$ and $b_0 = \cos\phi$ for some ϕ . If $a = \lim_{i\to\infty} a_i = \lim_{i\to\infty} b_i$, then

$$a = \frac{\pi}{2K(\phi)} , \tag{7.1}$$

where $K(\phi)$ is the complete elliptic integral of the first kind, i.e.,

$$K(\phi) = \int_0^{\pi/2} (1 - \sin^2\phi\sin^2\theta)^{-\frac{1}{2}}d\theta .$$

(A simple proof of (7.1) is given in Melzak [73].)

Also, if $c_0 = \sin\phi$, $c_{i+1} = a_i - a_{i+1}$ $(i=0,1,\ldots)$, then

$$\sum_{i=0}^{\infty} 2^{i-1}c_i^2 = 1 - \frac{E(\phi)}{K(\phi)} , \tag{7.2}$$

where $E(\phi)$ is the complete elliptic integral of the second kind, i.e.,

$$E(\phi) = \int_0^{\pi/2} (1 - \sin^2\phi\sin^2\theta)^{\frac{1}{2}}d\theta .$$

Both (7.1) and (7.2) were known by Gauss.

Legendre's Identity

For future use, we note the identity

$$K(\phi)E(\phi') + K(\phi')E(\phi) - K(\phi)K(\phi') = \frac{1}{2}\pi \ , \tag{7.3}$$

where $\phi + \phi' = \frac{1}{2}\pi$. (Legendre [11] proved by differentiation that the left side of (7.3) is constant, and the constant may be determined by letting $\phi \to 0$.)

8. FAST MULTIPLE-PRECISION EVALUATION OF π

The classical methods for evaluating π to precision n take time $O(n^2)$: see, for example, Shanks and Wrench [62]. Several methods which are asymptotically faster than $O(n^2)$ are known. For example, in Brent [75a] a method which requires time $O(M(n)\log^2(n))$ is described. From the bound (1.1) on $M(n)$, this is faster than $O(n^{1+\varepsilon})$ for any $\varepsilon > 0$.

Asymptotically the fastest known methods require time $O(M(n)\log(n))$. One such method is sketched in Beeler et al [72]. The method given here is faster, and does not require the preliminary computation of e .

The Gauss-Legendre method

Taking $\phi = \phi' = \pi/4$ in (7.3), and dividing both sides by π^2 , we obtain

$$[2K(\pi/4)E(\pi/4) - K^2(\pi/4)]/\pi^2 = \frac{1}{2\pi} \ . \tag{8.1}$$

However, from the A-G mean iteration with $a_0 = 1$ and $b_0 = 2^{-\frac{1}{2}}$, and the relations (7.1) and (7.2), we can evaluate $K(\pi/4)/\pi$ and $E(\pi/4)/\pi$, and thus the left side of (8.1). A division then gives π . (The idea of using (7.3) in this way occurred independently to Salamin [75] and Brent [75b].) After a little simplification, we obtain the following algorithm (written in pseudo-Algol):

$A \leftarrow 1;\quad B \leftarrow 2^{-\frac{1}{2}};\quad T \leftarrow 1/4;\quad X \leftarrow 1;$

while $A - B > 2^{-n}$ **do**

  begin $Y \leftarrow A;\quad A \leftarrow \frac{1}{2}(A + B);\quad B \leftarrow (BY)^{\frac{1}{2}};$

    $T \leftarrow T - X(A - Y)^2;$

    $X \leftarrow 2X$

  end;

return A^2/T [or, better, $(A + B)^2/(4T)$] .

The rate of convergence is illustrated in Table 8.1.

Table 8.1: Convergence of the Gauss-Legendre Method

Iteration	$A^2/T - \pi$	$\pi - (A + B)^2/(4T)$
0	8.6'-1	2.3'-1
1	4.6'-2	1.0'-3
2	8.8'-5	7.4'-9
3	3.1'-10	1.8'-19
4	3.7'-21	5.5'-41
5	5.5'-43	2.4'-84
6	1.2'-86	2.3'-171
7	5.8'-174	1.1'-345
8	1.3'-348	1.1'-694
9	6.9'-698	6.1'-1393

Since the A-G mean iteration converges with order 2, we need $\sim \log_2 n$ iterations to obtain precision n. Each iteration involves one (precision n) square root, one multiplication, one squaring, one multiplication by a power of two, and some additions. Thus, from the results of Section 2, the time required to evaluate π is $\sim \frac{15}{2} M(n) \log_2 n$.

Comments

1. Unlike Newton's iteration, the A-G mean iteration is not self-correcting. Thus, we cannot start with low precision

and increase it, as was possible in Section 2.

2. Since there are $\sim\log_2 n$ iterations, we may lose $O(\log\log(n))$ bits of accuracy through accumulation of rounding errors, even though the algorithm is numerically stable. Thus, it may be necessary to work with precision $n + O(\log\log(n))$. From (1.3), the time required is still $\sim\frac{15}{2}M(n)\log_2 n$.

9. MULTIPLE-PRECISION EVALUATION OF LOG(X)

There are several algorithms for evaluating $\log(x)$ to precision n in time $O(M(n)\log(n))$. For example, a method based on Landen transformations of incomplete elliptic integrals is described in Brent [75b]. The method described here is essentially due to Salamin (see Beeler et al [72]), though the basic relation (9.1) was known by Gauss.

If $\cos(\phi) = \varepsilon^{\frac{1}{2}}$ is small, then

$$K(\phi) = (1 + O(\varepsilon))\log(4\varepsilon^{-\frac{1}{2}}) \tag{9.1}$$

Thus, taking $a_0 = 1$, $b_0 = 4/y$, where $y = 4\varepsilon^{-\frac{1}{2}}$, and applying the A-G mean iteration to compute $a = \lim_{i\to\infty} a_i$, gives

$$\log(y) = \frac{\pi}{2a}(1 + O(y^{-2}))$$

for large y. Thus, so long as $y \geqslant 2^{n/2}$, we can evaluate $\log(y)$ to precision n. If $\log(y) = O(n)$ then $\sim 2\log_2 n$ iterations are required, so the time is $\sim 13M(n)\log_2 n$, assuming π is precomputed.

For example, to find $\log(10^6)$ we start the A-G mean iteration with $a_0 = 1$ and $b_0 = 4'-6$. Results of the first seven iterations are given to 10 significant figures in Table 9.1. We find that $\pi/(2a_7) = 13.81551056$, which is correct.

Table 9.1: Computation of $\log(10^6)$

i	a_i	b_i
0	1.000000000'0	4.000000000'-6
1	5.000020000'-1	2.000000000'-3
2	2.510010000'-1	3.162283985'-2
3	1.413119199'-1	8.909188753'-2
4	1.152019037'-1	1.122040359'-1
5	1.137029698'-1	1.136930893'-1
6	1.136980295'-1	1.136980294'-1
7	1.136980295'-1	1.136980295'-1

Since $\log(2) = \frac{1}{n}\log(2^n)$, we can evaluate $\log(2)$ to precision n in time $\sim 13M(n)\log_2 n$. Suppose $x \in [b,c]$, where $b > 1$. We may set $y = 2^n x$, evaluate $\log(y)$ as above, and use the identity

$$\log(x) = \log(y) - n.\log(2)$$

to evaluate $\log(x)$. Since $\log(y) \simeq n.\log(2)$, approximately $\log_2 n$ significant bits will be lost through cancellation, so it is necessary to work with precision $n + O(\log(n))$.

If x is very close to 1 , we have to be careful in order to obtain $\log(x)$ with a small relative error. Suppose $x = 1 + \delta$. If $|\delta| < 2^{-n/\log(n)}$ we may use the power series

$$\log(1 + \delta) = \delta - \delta^2/2 + \delta^3/3 - \dots ,$$

and it is sufficient to take about $\log(n)$ terms. If δ is larger, we may use the above A-G mean method, with working precision $n + O(n/\log(n))$ to compensate for any cancellation.

Finally, if $0 < x < 1$, we may use $\log(x) = -\log(1/x)$, where $\log(1/x)$ is computed as above. To summarize, we have proved:

Theorem 9.1

If $x > 0$ is a precision n number, then $\log(x)$ may be evaluated to precision n in time $\sim 13M(n)\log_2 n$ as $n \to \infty$ [assuming π and $\log(2)$ precomputed to precision $n + O(n/\log(n))$].

Note: The time required to compute $\log(x)$ by the obvious power series method is $O(nM(n))$. Since $13\log_2 n < n$ for $n \geq 83$, the method described here may be useful for moderate n, even if the classical $O(n^2)$ multiplication algorithm is used.

10. MULTIPLE-PRECISION EVALUATION OF EXP(X)

Corresponding to Theorem 9.1, we have:

Theorem 10.1

If $[a,b]$ is a fixed interval, and $x \in [a,b]$ is a precision n number such that $\exp(x)$ does not underflow or overflow, then $\exp(x)$ can be evaluated to precision n in time $\sim 13M(n)\log_2 n$ as $n \to \infty$ (assuming π and $\log(2)$ are precomputed).

Proof

To evaluate $\exp(x)$ we need to solve the equation $f(y) = 0$, where $f(y) = \log(y) - x$, and x is regarded as constant. Since

$$f^{(k)}(y) = (-1)^{k-1}(k-1)!y^{-k}$$

can be evaluated in time $O(M(n)) = o(M(n)\log(n))$ for any fixed $k \geq 1$, the result follows from Theorems 6.1 and 9.1. [The $(k + 1)$-th order method in the proof of Theorem 6.1 may simply be taken as

$$y_{i+1} = y_i \sum_{j=0}^{k} (x - \log(y_i))^j/j! \ ]$$

11. MULTIPLE-PRECISION OPERATIONS ON COMPLEX NUMBERS

Before considering the multiple-precision evaluation of trigonometric functions, we need to state some results on multiple-precision operations with complex numbers. We assume that a precision n complex number $z = x + iy$ is represented as a pair (x, y) of precision n real numbers. As before, a precision n operation is one which gives a result with a relative error $O(2^{-n})$. (Now, of course, the relative error may be complex, but its absolute value must be $O(2^{-n})$.) Note that the smaller component of a complex result may occasionally have a large relative error, or even the wrong sign!

Complex Multiplication

Since $z = (t + iu)(v + iw) = (tv - uw) + i(tw + uv)$, a complex multiplication may be done with four real multiplications and two additions. However, we may use an idea of Karatsuba and Ofman [62] to reduce the work required to three real multiplications and some additions: evaluate tv, uw, and $(t + u)(v + w)$, then use

$$tw + uv = (t + u)(v + w) - (tv + uw) .$$

Since $|t + u| \leq 2^{\frac{1}{2}}|t + iu|$ and $|v + w| \leq 2^{\frac{1}{2}}|v + iw|$, we have

$$|(t + u)(v + w)| \leq 2|z| .$$

Thus, all rounding errors are of order $2^{-n}|z|$ or less, and the computed product has a relative error $O(2^{-n})$. The time for the six additions is asymptotically negligible compared to that for the three multiplications, so precision n complex multiplication may be performed in time $\sim 3M(n)$.

Complex Squares

Since $(v + iw)^2 = (v - w)(v + w) + 2ivw$, a complex

square may be evaluated with two real multiplications and additions, in time $\sim 2M(n)$.

Complex Division

Using complex multiplication as above, and the same division algorithm as in the real case, we can perform complex division in time $\sim 12M(n)$. However, it is faster to use the identity

$$\frac{t + iu}{v + iw} = (v^2 + w^2)^{-1}[(t + iu)(v - iw)] ,$$

reducing the problem to one complex multiplication, four real multiplications, one real reciprocal, and some additions. This gives time $\sim 10M(n)$. For complex reciprocals we have $t = 1$, $u = 0$, and time $\sim 7M(n)$.

Complex Square Roots

Using (2.2) requires, at the last iteration, one precision n complex squaring and one precision n/2 complex division. Thus, the time required is $\sim 2(2 + 10/2)M(n) = 14M(n)$.

Complex A-G Mean Iteration

From the above results, a complex square root and multiplication may be performed in time $\sim 17M(n)$. Each iteration transforms two points in the complex plane into two new points, and has an interesting geometric interpretation.

12. MULTIPLE-PRECISION EVALUATION OF TRIGONOMETRIC FUNCTIONS

Since

(12.1) $$\log(v + iw) = \log|v + iw| + i.\mathrm{artan}(w/v)$$

and

(12.2) $$\exp(i\theta) = \cos(\theta) + i.\sin(\theta) ,$$

we can evaluate $\mathrm{artan}(x)$, $\cos(x)$ and $\sin(x)$ if we can evaluate $\log(z)$ and $\exp(z)$ for complex arguments z . This may be done just as described above for real z, provided we choose the correct value of $(a_j b_j)^{\frac{1}{2}}$. Some care is necessary to avoid excessive cancellation; for example, we should use the power series for $\sin(x)$ if $|x|$ is very small, as described above for $\log(1 + \delta)$. Since $\sim 2\log_2 n$ A-G mean iterations are required to evaluate $\log(z)$, and each iteration requires time $\sim 17M(n)$, we can evaluate $\log(z)$ in time $\sim 34M(n)\log_2 n$. From the complex version of Theorem 6.1, $\exp(z)$ may also be evaluated in time $\sim 34M(n)\log_2 n$.

As an example, consider the evaluation of $\log(z)$ for $z = 10^6(2 + i)$. The A-G mean iteration is started with $a_0 = 1$ and $b_0 = 4/z = 1.6'-6 - (8.0'-7)i$. The results of six iterations are given, to 8 significant figures, in Table 12.1.

Table 12.1: Evaluation of $\log 10^6(2 + i)$.

j	a_j	b_j
0	(1.0000000'0, 0.0000000'0)	(1.6000000'-6, -8.0000000'-7)
1	(5.0000080'-1, -4.0000000'-7)	(1.3017017'-3, -3.0729008'-4)
2	(2.5065125'-1, -1.5384504'-4)	(2.5686505'-2, -2.9907884'-3)
3	(1.3816888'-1, -1.5723167'-3)	(8.0373334'-2, -4.6881008'-3)
4	(1.0927111'-1, -3.1302088'-3)	(1.0540970'-1, -3.6719673'-3)
5	(1.0734040'-1, -3.4010880'-3)	(1.0732355'-1, -3.4064951'-3)
6	(1.0733198'-1, -3.4037916'-3)	(1.0733198'-1, -3.4037918'-3)

We find that
$$\frac{\pi}{2a_7} = 14.620230 + 0.46364761i$$
$$\simeq \log|z| + i.\text{artan}(\tfrac{1}{2})$$

as expected.

Another method for evaluating trigonometric functions in time $O(M(n)\log(n))$, without using the identities (12.1) and (12.2), is described in Brent [75b].

13. OPERATIONS ON FORMAL POWER SERIES

There is an obvious similarity between a multiple-precision number with base β :

$$\beta^e \sum_{i=1}^{n} a_i \beta^{-i} \quad (0 \leq a_i < \beta) ,$$

and a formal power series:

$$\sum_{i=0}^{\infty} a_i x^i \qquad (a_i \text{ real, } x \text{ an indeterminate}) .$$

Thus, it is not surprising that algorithms similar to those described in Section 2 may be used to perform operations on power series.

In this section only, $M(n)$ denotes the number of scalar operations required to evaluate the first n coefficients $c_0,\ldots,c_{n-1}$ in the formal product

$$\left(\sum_{i=0}^{\infty} a_i x^i\right) \left(\sum_{i=0}^{\infty} b_i x^i\right) = \sum_{i=0}^{\infty} c_i x^i .$$

Clearly, c_j depends only on $a_0,\ldots,a_j$ and $b_0,\ldots,b_j$, in fact

$$c_j = \sum_{i=0}^{j} a_i b_{j-i} .$$

The classical algorithm gives $M(n) = O(n^2)$, but it is possible to use the fast Fourier transform (FFT) to obtain

$$M(n) = O(n.\log(n)) .$$

(see Borodin [73]).

If we assume that $M(n)$ satisfies conditions (1.2) and

(1.3), then the time bounds given in Section 2 for division, square roots, etc. of multiple-precision numbers also apply for the corresponding operations on power series (where we want the first n terms in the result). For example, if $P(x) = \sum_{i=0}^{\infty} a_i x^i$ and $a_0 \neq 0$, then the first n terms in the expansion of $1/P(x)$ may be found with $\sim 3M(n)$ operations as $n \to \infty$. However, some operations, e.g. computing exponentials, are much easier for power series than for multiple-precision numbers!

Evaluation of $\log(P(x))$

If $a_0 > 0$ we may want to compute the first n terms in the power series $Q(x) = \log(P(x))$. Since $Q(x) = \log(a_0) + \log(P(x)/a_0)$, there is no loss of generality in assuming that $a_0 = 1$. Suppose $Q(x) = \sum_{i=0}^{\infty} b_i x^i$. From the relation

$$Q'(x) = P'(x)/P(x) , \tag{13.1}$$

where the prime denotes formal differentiation with respect to x, we have

$$\sum_{i=1}^{\infty} i b_i x^{i-1} = \left(\sum_{i=1}^{\infty} i a_i x^{i-1}\right) \bigg/ \left(\sum_{i=0}^{\infty} a_i x^i\right) . \tag{13.2}$$

The first n terms in the power series for the right side of (13.2) may be evaluated with $\sim 4M(n)$ operations, and then we need only compare coefficients to find $b_1, \ldots, b_{n-1}$. (Since $a_0 = 1$, we know that $b_0 = 0$.) Thus, the first n terms in $\log(P(x))$ may be found in $\sim 4M(n)$ operations. It is interesting to compare this result with Theorem 9.1.

Evaluation of $\exp(P(x))$

If $R(x) = \exp(P(x))$ then $R(x) = \exp(a_0)\exp(P(x) - a_0)$, so there is no loss of generality in assuming that $a_0 = 0$. Now $\log(R(x)) - P(x) = 0$, and we may regard this as an equation for the unknown power series $R(x)$, and solve it by one of the usual iterative methods. For example, Newton's method gives the iteration

$$R_{i+1}(x) = R_i(x) - R_i(x)(\log(R_i(x)) - P(x)) . \tag{13.3}$$

If we use the starting approximation $R_0(x) = 1$, then the terms in $R_k(x)$ agree exactly with those in $R(x)$ up to (but excluding) the term $O(x^{2^k})$. Thus, using (13.3), we can find the first n terms of $\exp(P(x))$ in $\sim 9M(n)$ operations, and it is possible to reduce this to $\sim\frac{22}{3}M(n)$ operations by using a fourth-order method instead of (13.3). Compare Theorem 10.1.

Evaluation of P^m

Suppose we want to evaluate $(P(x))^m$ for some large positive integer m . We can assume that $a_0 \neq 0$, for otherwise some power of x may be factored out. Also, since $P^m = a_0^m(P/a_0)^m$, we can assume that $a_0 = 1$. By forming P^2 , P^4 , P^8 , ..., and then the appropriate product given by the binary expansion of m , we can find the first n terms of P^m in $O(M(n)\log_2 m)$ operations. Surprisingly, this is not the best possible result, at least for large m . From the identity

$$P^m = \exp(m.\log(P)) \tag{13.4}$$

and the above results, we can find the first n terms of P^m in $O(M(n))$ operations! (If $a_0 \neq 1$, we also need $O(\log_2 m)$ operations to evaluate a_0^m .) If the methods described above are used to compute the exponential and logarithm in (13.4), then the number of operations is $\sim\frac{34}{3}M(n)$ as $n \to \infty$.

Other Operations on Power Series

The method used to evaluate $\log(P(x))$ can easily be generalized to give a method for $f(P(x))$, where $df(t)/dt$

is a function of t which may be written in terms of square roots, reciprocals etc. For example, with $f(t) = \text{artan}(t)$ we have $df/dt = 1/(1 + t^2)$, so it is easy to evaluate $\text{artan}(P(x))$. Using Newton's method we can evaluate the inverse function $f^{(-1)}(P(x))$ if $f(P(x))$ can be evaluated. Generalizations and applications are given in Brent and Kung [75].

Some operations on formal power series do not correspond to natural operations on multiple-precision numbers. One example, already mentioned above, is formal differentiation. Other interesting examples are composition and reversion. The classical composition and reversion algorithms, as given in Knuth [69], are $O(n^3)$, but much faster algorithms exist: see Brent and Kung [75].

REFERENCES

Beeler, Gosper and Schroeppel [72] Beeler, M., Gosper, R.W., and Schroeppel, R. "Hakmem". Memo No. 239, M.I.T. Artificial Intelligence Lab., 1972, 70-71.

Borodin [73] Borodin, A., "On the number of arithmetics required to compute certain functions - circa May 1973". In Complexity of Sequential and Parallel Numerical Algorithms (ed. by J.F. Traub), Academic Press, New York, 1973, 149-180.

Brent [75a] Brent, R.P., "The complexity of multiple-precision arithmetic". Proc. Seminar on Complexity of Computational Problem Solving (held at the Australian National University, Dec. 1974), Queensland Univ. Press, Brisbane, 1975.

Brent [75b] Brent, R.P., "Fast multiple-precision evaluation of elementary functions". Submitted to J. ACM.

Brent and Kung [75] Brent, R.P. and Kung, H.T., "Fast algorithms for reversion and composition of power series". To appear. (A preliminary paper appears in these Proceedings.)

Gauss [1876] Gauss, C. F., "Carl Friedrich Gauss Werke", (Bd. 3), Göttingen, 1876, 362-403.

Karatsuba and Ofman [62] Karatsuba, A. and Ofman, Y., "Multiplication of multidigit numbers on automata", (in Russian). Dokl. Akad. Nauk SSSR 146 (1962), 293-294.

Knuth [69] Knuth, D.E., "The Art of Computer Programming", (Vol. 2), Addison Wesley, Reading, Mass., 1969, Sec. 4.7.

Legendre [11] Legendre, A.M., "Exercices de Calcul Integral", (Vol. 1), Paris, 1811, 61.

Melzak [73] Melzak, Z.A., "Companion to Concrete Mathematics", Wiley, New York, 1973, 68-69.

Salamin [75] Salamin, E., "A fast algorithm for the computation of π". To appear in Math. Comp.

Schönhage and Strassen [71] Schönhage, A. and Strassen, V., "Schnelle Multiplikation grosser Zahlen". Computing 7 (1971), 281-292.

Shanks and Wrench [62] Shanks, D. and Wrench, J.W., "Calculation of π to 100,000 decimals". Math. Comp. 16 (1962), 76-99.

NUMERICAL STABILITY OF ITERATIONS FOR SOLUTION OF NONLINEAR EQUATIONS AND LARGE LINEAR SYSTEMS

H. Woźniakowski
Department of Computer Science
Carnegie-Mellon University
(On leave from University of Warsaw)

ABSTRACT

We survey some recent results on the problem of numerical stability of iterations for the solution of nonlinear equations $F(x) = 0$ and large linear systems $Ax+g = 0$ where $A = A^*$ is positive definite.

For systems of nonlinear equations we assume that the function F depends on a so called data vector $F(x) = F(x;d)$. We define the condition number cond$(F;d)$, numerical stability and well-behavior of iterations for the solution of $F(x) = 0$. Necessary and sufficient conditions for a stationary iteration to be numerically stable and well-behaved are presented. We show that Newton iteration for the multivariate case and secant iteration for the scalar case are well-behaved.

For large linear systems we present the rounding error analysis for the Chebyshev iteration and for the successive approximation iterations. We show that these iterations are numerically stable and that the condition number of A is a crucial parameter.

1. INTRODUCTION

Any iterative algorithm for the solution of nonlinear equations or large linear systems should satisfy a number of criteria such as good convergence properties, numerical stability and as small complexity as possible. Since any iteration is implemented in floating point arithmetic, due to rounding errors one can at best count on approximate properties of this iteration.

In this paper we survey some recent results on the problem of numerical stability of iterations for solving nonlinear equations or large linear systems. Section 2, which deals with numerical stability of iterations for nonlinear equations, is primarily based on the author's paper [75a]. Section 3, which deals with numerical stability of iterations for large linear systems, is based on the author's papers [75b] and [75d].

It might seem that the problem of numerical stability of iterations is not as important as for direct methods. We show that the condition number of the problem is crucial and if the problem is ill-conditioned, then it is impossible to compute a good approximation of the solution no matter how sophisticated an iteration is used. Furthermore, if the problem is well-conditioned, then we can compute a good approximation whenever the iteration used is numerically stable.

2. NUMERICAL STABILITY OF ITERATIONS FOR NONLINEAR EQUATIONS

We approximate a simple zero α of the nonlinear function F,

(2.1) $F(x) = 0$

where $F: D \subset \mathbb{C}^N \to \mathbb{C}^N$ and $\mathbb{C}^N$ is the N dimensional complex space. Throughout this section we assume that F depends parametrically on a vector d which will be called a <u>data vector</u>, $F(x) = F(x;d)$, and $d \in \mathbb{C}^m$. For many problems d is given explicitly, e.g., $F(x) = \sum_{i=0}^{n} a_i x^i$ for $N = 1$. For certain F it is not obvious how to define d, e.g. $F(x) = x^2 - e^x$.

One idea how to determine d is as follows. We solve (2.1) by iteration and most practical iterations use the value of F(x) to get the next approximation to α. We compute F(x;d) in floating point binary arithmetic (fl), see Wilkinson [63], and at best we can expect that a slightly perturbed computed value fl(F(x;d)) is the exact one for a slightly perturbed function at slightly perturbed inputs (see Kahan [71]), i.e.,

$$(2.2) \quad fl(F(x;d)) = (I-\Delta F)F(x+\Delta x;d+\Delta d)$$

for $||\Delta F|| \le K_1\ 2^{-t}||I||$, $||\Delta x|| \le K_2\ 2^{-t}||x||$ and $||\Delta d|| \le K_3\ 2^{-t}||d||$ where $K_i = K_i(N;m)$ and 2^{-t} is the relative computer precision.

The condition (2.2) can be treated as an equation on a data vector.

We have to represent the data vector d in fl. Let $\tilde{d} = rd(d)$ denote t digit representation of d in fl. Then

$$(2.3) \quad ||\tilde{d}-d|| \le K_4 2^{-t}||d|| \qquad \text{where } K_4 = K_4(m).$$

Due to this unavoidable change of the data vector instead of the problem $F(x;d) = 0$ we can at best approximate a solution of the problem $F(x;\tilde{d}) = 0$. Let $\tilde{\alpha}$ be a simple zero of $F(x;\tilde{d}) = 0$. It is easy to verify that for sufficiently smooth F we get

$$(2.4) \quad \tilde{\alpha} - \alpha = -F_x'(\alpha;d)^{-1}\ F_d'(\alpha;d)(\tilde{d}-d) + O(2^{-2t})$$

where F_x' and F_d' denote the first derivative with respect to x and d. For $\alpha \neq 0$ we have

$$(2.5) \quad \frac{\|\tilde{\alpha}-\alpha\|}{\|\alpha\|} \le K_4 2^{-t} \text{ cond}(F;d) + O(2^{-2t})$$

where

$$(2.6) \quad \text{cond}(F;d) = \|F'_x(\alpha;d)^{-1} F'_d(\alpha;d)\| \frac{\|d\|}{\|\alpha\|}$$

is called the condition number of F with respect to the data vector d.

The condition number measures the relative sensitivity of the solution with respect to a small relative perturbation of the data vector.

Note that in general cond(F;d) is not related to the condition number $H(F'(\alpha))$ of the first derivative $F'(\alpha)$, $H(F'(\alpha)) = \|F'(\alpha)\| \, \|F(\alpha)^{-1}\|$ which occurs in linear analysis.

Having the concept of the condition number we define numerical stability and well-behavior of iterations for the solution of $F(x;d) = 0$.

Let $\{x_k\}$ be a computed sequence of the successive approximations of α by an iteration φ in fl.

An iteration φ is called numerically stable if

$$(2.7) \quad \overline{\lim_k} \frac{\|x_k-\alpha\|}{\|\alpha\|} \le 2^{-t}(K_5 + K_6 \text{ cond}(F;d)) + O(2^{-2t})$$

where $K_i = K_i(N,m)$ for $i = 5,6$.

An iteration φ is called well-behaved if

$$(2.8) \quad \overline{\lim_k} \|F(x_k+\delta x_k;\ d+\delta d_k)\| = O(2^{-2t})$$

where $\|\delta x_k\| \le K_7 \, 2^{-t}\|x_k\|$, $\|\delta d_k\| \le K_8 \, 2^{-t}\|d\|$.

Numerical stability states that the relative error of the computed x_k is of order $2^{-t}\text{cond}(F;d)$. Well-behavior states that a slightly perturbed computed x_k, k large, is an almost exact solution of a slightly perturbed problem.

Note that if φ is well-behaved, then φ is numerically stable but not vice versa except the scalar case $N = 1$ (see Lemma 4.1 in Woźniakowski [75a]).

Assume that φ is a stationary iteration which produces in exact arithmetic the next approximation x_{k+1} equal to

$$(2.9)\quad x_{k+1}^{*} = \varphi(x_k,\ldots,x_{k-n}, \mathfrak{N}(x_k,\ldots,x_{k-n},F))$$

where n denotes the size of the iteration memory (see Traub [64]) and $\mathfrak{N}$ is information of F at $x_k,\ldots,x_{k-n}$. Next suppose that

$$\|x_{k+1}^{*}-\alpha\| \le C \prod_{j=0}^{n} \|x_{k-j}-\alpha\|^{P_j}$$

where $P_j \ge 0$, $\sum_{j=0}^{n} P_j \ge 2$ and $C = C(F)$ whenever $\|x_k-\alpha\|\le\ldots\le \|x_{k-n}-\alpha\|\le \Gamma$ for sufficiently small Γ.

In floating point arithmetic instead of (2.9) we have

$$(2.10)\quad x_{k+1} = x_{k+1}^{*} + \xi_k$$

where ξ_k is the computer error in one iterative step. The value of ξ_k depends mainly on the computed error of the information $\mathfrak{N}$ and on the computed error of an algorithm which is used to perform one iterative step.

It is possible to find a form of ξ_k to ensure numerical stability and well-behavior of the stationary iteration φ. Namely, φ is numerically stable iff

$$(2.11)\quad \overline{\lim_{k}} \frac{\|\xi_k\|}{\|\alpha\|} \le 2^{-t}(K_9 + K_{10}\,\mathrm{cond}(F;d)) + O(2^{-2t})$$

where $K_i = K_i(N,m)$ for $i = 9,10$, and φ is well-behaved iff

$$(2.12)\quad \xi_k = \Delta x_k + F_x'(x_k;d)^{-1}\, F_d'(x_k;d)\Delta d_k + O(2^{-2t})$$

where $\|\Delta x_k\| \le K_{11}2^{-t}\|x_k\|$, $\|\Delta d_k\| \le K_{12}2^{-t}\|d\|$ for large k and

$K_i = K_i(N,m)$, $i = 11$ and 12. (See Theorem 4.1 and Corollary 4.2 in Woźniakowski [75a].)

Using (2.11) and (2.12) one can verify that Newton iteration is well-behaved under the following assumptions:

(i) $F(x_k;d)$ is computed by a well-behaved algorithm (see (2.2))

(ii) $fl(F'(x_k;d)) = F'(x_k) + O(2^{-t})$

(iii) the computed z_k, ($z_k = -F'(x_k)^{-1}F(x_k)$ and $x_{k+1} = x_k + z_k$) satisfies $(fl(F'(x_k;d))+E_k)z_k = -fl(F(x_k;d))$, $E_k = O(2^{-t})$.

The first two conditions require a certain accuracy in $F(x_k)$ and $F'(x_k)$ whereas the last conditions mean that z_k is the exact solution of a perturbed system which holds if Gaussian elimination with pivoting or the Householder method is used.

An interesting question is whether the secant iteration is well-behaved. For the scalar case secant iteration produces

$$x_{k+1}^{*} = x_k - \frac{x_k - y_k}{F(x_k) - F(y_k)} F(x_k)$$

where $y_k = x_{k-1}$ (with memory) or $y_k = x_k + \gamma_k F(x_k)$ (two-point iteration) for a certain γ_k.

It is shown in Woźniakowski [75a] that secant iteration is well-behaved whenever

$$(2.13) \qquad \left| \frac{F(x_k)}{F(x_k) - F(y_k)} \right| \le Q$$

for all $k \ge k_0$ and a positive constant Q independent of F. Note that (2.13) does not hold for the Steffenson iteration, $y_k = x_k + F(x_k)$. It may be shown that with this choice of

y_k, secant iteration is unstable. For secant iteration with memory

$$\frac{F(x_k)}{F(x_k)-F(x_{k-1})} \cong 0(x_{k-2}-\alpha) + 0\left(\frac{2^{-t}}{x_{k-1}-\alpha}\right).$$

Thus (2.13) holds as long as $|x_{k-1}-\alpha| \gg 2^{-t}$.

Numerical stability of the multivariate secant iteration was proved by Jankowska [75] under some assumptions on a suitable distance and position of successive approximations. Well-behavior of the multivariate secant iteration is open.

There are several classes of iterations of practical interest for which the problem of numerical stability is open. Examples are interpolatory iterations $I_{n,s}$ for the scalar case and $I_{0,s}$ for the multivariate one, integral-interpolatory iterations $I_{-1,s}$ and hermitian multipoint iterations (see Traub [64], Kacewicz [75a] and [75b], Kung and Traub [74] and Woźniakowski [75c] respectively). One interesting question is how to use iterations with memory in a stable way. There are some reasons to believe that at least some of the mentioned classes of iterations are numerically stable under certain assumptions but further research is needed.

3. NUMERICAL STABILITY OF ITERATIONS FOR LARGE LINEAR SYSTEMS

Direct methods of numerical interest for the solution of linear systems $Ax+g = 0$ where A is $N \times N$ matrix and g is $N \times 1$ vector are <u>well-behaved</u>. Specifically they produce an approximation y to the exact solution α such that y is the exact solution for a slightly perturbed A,

(3.1) $(A+E)y + g = 0$

where $\|E\| \le c_1 2^{-t}\|A\|$ and $c_1 = c_1(N)$.

Examples of well-behaved direct methods include Gaussian elimination with pivoting, the Householder method and the Gram-Schmidt reorthogonalization method. Note that a method is well-behaved iff the residual vector $r = Ay+g$ is small, i.e.

$$(3.2) \quad \|r\| \le C_2\, 2^{-t}\|A\|\,\|y\|, \qquad C_2 = C_2(N).$$

Furthermore, for any well-behaved method we get

$$(3.3) \quad \frac{\|y-\alpha\|}{\|\alpha\|} \le C_3\, 2^{-t}\, H(A)$$

where $H(A) = \|A\|\,\|A^{-1}\|$ denotes the condition number of A and $C_3 = C_3(N)$. In general (3.3) is sharp which indicates that the condition number H(A) is a crucial parameter. (Note that (3.3) also holds for any numerically stable method.)

It might seem that the numerical accuracy of iterations for solving large linear systems might be better than for direct methods. However, this is not true. We shall discuss some iterations to see that H(A) is still crucial and moreover, we shall show that for some very efficient iterations well-behavior does not hold in general.

Two reasons why the condition number is still crucial are as follows:

(i) No matter which iteration is used we have to represent (not necessarily store!) all entries of A and g in floating point arithmetic. Thus, instead of the problem $Ax+g = 0$ we can at best approximate the solution $\tilde{\alpha}$ of

$$(3.4) \qquad (A+\delta A)x + (g+\delta g) = 0$$

where $\|\delta A\| \le C_4\, 2^{-t}\|A\|$, $\|\delta g\| \le C_5\, 2^{-t}\|g\|$ and

and $C_i = C_i(N)$ for $i = 4,5$. The relative error $\|\tilde{\alpha}-\alpha\|/\|\alpha\|$ is of order $\underline{2^{-t}H(A)}$. Thus, once more the condition number is important.

(ii) Let us assume that all entries of A and g can be exactly represented in fl, $A = rd(A)$, $g = rd(g)$. For many iterations the only known information of the system is given by a procedure which for a given x computes $z = Ax$. Since Ax is computed in fl then at best we can get

$$z = fl(Ax) = (A+E)x \tag{3.5}$$

where $E = E(x)$ and $\|E\| \leq C_5\, 2^{-t}\|A\|$, $C_5 = C_5(N)$. Thus all information derives from perturbed systems and the computed solution x_k can be at best the exact solution of a slightly perturb problem $(A+E_k) + g = 0$. Then

$$x_k - \alpha = -A^{-1}E_k x_k. \tag{3.6}$$

As long as we do not require any special property of E_k then $\|A^{-1}E_k x_k\|$ can be close to $\|A^{-1}\|\|E\|\|x_k\|$ which is of order $2^{-t}H(A)\|x_k\|$.

We are now in a position to discuss numerical properties of some particular iterations. First we consider successive approximation iterations which are defined as follows:

(i) Transform $Ax+g = 0$ to the equivalent system

$$x = Bx + h. \tag{3.7}$$

Sometimes B is chosen to minimize the spectral radius $\rho(B)$ of B, $\rho(B) < 1$.

(ii) Solve (3.7) by the iteration

(3.8) $$x_{k+1} = Bx_k + h, \quad k = 0,1,\ldots$$

where x_0 is a given initial approximation.

For different transformations we get different iterations; for instance, the Jacobi (J), Richardson (R), Gauss-Seidel (GS) or successive overrelaxation (SOR) iterations, see Young [71]. Note that for $e_k = x_k - \alpha$ we get $e_k = B^k e_0$ and the character of convergence mainly depends on the spectral radius $\rho(B)$.

Suppose that in fl we have

(3.9) $$fl(Bx_k+h) = (B+E_k)x_k + (I+\delta I_k)h = Bx_k+h + \xi_k$$

where $\|E_k\| \le C_6\, 2^{-t}\|B\|$, $\|\delta I_k\| \le C_7\, 2^{-t}$ and $C_i = C_i(N)$, $i = 6,7$ and

$$\xi_k = E_k x_k + \delta I_k (I-B)\alpha.$$

Thus, instead of (3.8) we get in fl,

$$x_{k+1} = Bx_k+h + \xi_k$$

which has the solution

(3.10) $$x_{k+1} - \alpha = B^{k+1}(x_0-\alpha) + \sum_{i=0}^{k} B^{k-i}\xi_i .$$

Suppose that $\|B\| < 1$. From (3.3) we get

$$\overline{\lim_k}\|x_k-\alpha\| \le \frac{1}{1-\|B\|}\overline{\lim_k}\|\xi_k\| \le C_8\, 2^{-t}\,\frac{\|B\|+\|I-B\|}{1-\|B\|}\,\|\alpha\|+O(2^{-2t})$$

where $C_8 = \max(C_6,C_7)$. Hence if

(3.11) $q = (\|B\| + \|I-B\|)/(1-\|B\|)$ is of order $\|A\|\,\|A^{-1}\|$

then this iteration is <u>numerically stable</u>.

For instance, for the Richardson iteration we get

$$B = I - cA \quad \text{where } A = A^* > 0$$

and $c = \dfrac{2}{\lambda_1+\lambda_2}$ for $\lambda_1 = \|A^{-1}\|_2^{-1}$, $\lambda_2 = \|A\|_2$.

Then $\|B\|_2 = (\lambda_2-\lambda_1)/(\lambda_2+\lambda_1)$ and

$$q = \frac{3}{2} H(A) - 1$$

which proves that the Richardson iteration is stable. (For more examples see Woźniakowski [75d].) However, it is very easy to find a counter example where (3.11) does not hold even for N = 1. (Note that for N = 1, H(A) = 1.) Let us consider

$$(2-c)x = 1 \qquad \text{for } 0 < c < 1$$

with the transformation $x = (-1+c)x + 1$. Thus $B = -1+c$ and $q = q(c) = \frac{3-2c}{c}$. Note that $\lim_{c\to 0+} q(c) = +\infty$ which indicates that for small c (3.11) does not hold. Numerical tests confirm this observation. For instance using the PDP-10 where $2^{-t} \doteq 10^{-8}$ with $c = 10^{-4}$ we get x_k such that $|x_k-\alpha|/|\alpha| \doteq 10^{-4}$.

It is possible to prove that if B is diagonalizable and there exists a constant k independent on B such that $|1-\lambda| \le k(1-|\lambda|)$ for all eigenvalues λ of B then (3.8) is well behaved (Stęwart [73] and Woźniakowski [75d]).

We pass to the second class of iterations for large linear systems $Ax+g = 0$ where $A = A^*$ is positive definite. We construct a sequence $\{x_k\}$ of the successive approximation of α such that

$$(3.12) \quad x_k - \alpha = W_k(A)(x_0-\alpha)$$

where $W_k(0) = 1$ and W_k is a polynomial of degree at most k.

In the Chebyshev iteration W_k is defined by

$$(3.13) \quad \|W_k\| = \inf_{P \in P_k(0,1)} \|P\|$$

where $\|P\| = \sup_{a \le x \le b} |P(x)|$, $P_k(0,1)$ denotes a class of polynomials of degree at most k which has the value 1 at zero and $[a,b]$ contains all eigenvalues of matrix A.

The solution of (3.13) is given by the Chebyshev polynomials of the first kind and using the three-terms recurrence formula we get

$$(3.14) \quad x_{k+1} = x_k + \left\{p_{k-1}(x_k - x_{k-1}) - r_k\right\}/q_k, \; r_k = Ax_k + g$$

for certain coefficients p_{k-1} and q_k.

Assuming that

$$(3.15) \quad fl(Ax) = (A+E)x, \; \|E\|_2 \le C_9 \, 2^{-t}\|A\|_2$$

and $a = \|A^{-1}\|_2^{-1}$, $b = \|A\|_2$ it is possible to show that the computed sequence $\{x_k\}$ by the Chebyshev iteration satisifes

$$(3.16) \quad \overline{\lim_k} \, \|x_k - \alpha\| \le (1+4C_9)2^{-t} \, H(A) + O(2^{-2t})$$

which means numerical stability. Unfortunately the Chebyshev iteration is, in general, not well-behaved since the computed residual vector r_k can be of order $2^{-t}\|A\| \, \|\alpha\| \, H(A)$, see Woźniakowski [75b]. It seems to us that for any numerically stable iteration based on (3.12), the norm $\max_k \|W_k\|$ has to be relatively small. Note that in conjugate gradient iterations W_k is defined as the polynomial which minimizes a certain norm of $x_k - \alpha$, see Stiefel [58]. This need not imply that $\|W_k\|$ is small. This might explain why conjugate gradiant iterations are numerically unstable.

ACKNOWLEDGMENT

I wish to thank J. F. Traub and B. Kacewicz for their comments on this paper.

REFERENCES

Kacewicz [75a] Kacewicz, B., "An Integral-Interpolatory Iterative Method for the Solution of Nonlinear Scalar Equations," Department of Computer Science Report, Carnegie-Mellon University, 1975.

Kacewicz [75b] Kacewicz, B., "The Use of Integrals in the Solution of Nonlinear Equations in N Dimensions," these **Proceedings**. Also, Department of Computer Science Report, Carnegie-Mellon University, 1975.

Kahan [71] Kahan, W., "A Survey of Error Analysis," IFIP Congress 1971, I, 220-226.

Kung and Traub [74] Kung, H. T. and J. F. Traub, "Optimal Order of One-Point and Multipoint Iteration," J. Assoc. Comput. Mach., Vol. 21, No. 4, 1974, 643-651.

Jankowska [75] Jankowska, J., "Numerical Analysis of Multivariate Secant Method," a part of the Ph.D. dissertation, University of Warsaw, 1975.

Stewart [73] Stewart, G. W., private communication.

Stiefel [58] Stiefel, E., "Kernel Polynomials in Linear Algebra and Their Numerical Applications," NBS Appl. Math., Series 40, 1958, 1-22.

Traub [64] Traub, J. F., Iterative Methods for the Solution of Equations, Prentice-Hall, Englewood Cliffs, New Jersey, 1964.

Wilkinson [63] Wilkinson, J. H., Rounding Errors in Algebraic Processes, Prentice-Hall, Englewood Cliffs, New Jersey, 1963.

Woźniakowski [75a] Woźniakowski, H., "Numerical Stability for Solving Nonlinear Equations," Department of Computer Science Report, Carnegie-Mellon University, 1975.

Woźniakowski [75b] Woźniakowski, H., "Numerical Stability of the Chebyshev Method for the Solution of Large Linear Systems," Department of Computer Science Report, Carnegie-Mellon University, 1975.

Woźniakowski [75c] Woźniakowski, H., "Maximal Order of Multipoint Iterations Using n Evaluations, " Department of Computer Science Report, Carnegie-Mellon University, 1975, these Proceedings.

Woźniakowski [75d] Woźniakowski, H., "Numerical Stability of the Successive Approximation Method for the Solution of Large Linear and Nonlinear Equations, in progress.

Young [71] Young, D. M., Iterative Solution of Large Linear Systems, Academic Press, New York, 1971.

ON THE COMPUTATIONAL COMPLEXITY OF APPROXIMATION OPERATORS II

JOHN R. RICE*

Purdue University

1. INTRODUCTION

Computational complexity is a measure of the number of operations that some abstract machine requires to carry out a task. The task considered here is to compute an approximation to a real function f(x) and the <u>only</u> operations that we count are evaluations of f(x). Thus, we consider all other arithmetic performed to be negligible. We have already considered this topic in a previous paper (see Rice, [73]), but we recast the terminology and notation to be more natural. We also sharpen many of the results of Rice [73] and establish some new results.

We consider approximation by polynomials and piecewise polynomials in some norm (primarily L_2 and L_∞). For a given number N of parameters (coefficients or knots) let $P_N^*(x)$ denote the best approximation and let $\varepsilon(N)$ denote its error $||f-P_N^*||$. Throughout we assume the approximation is on a standard interval. Note that P_N^* and $\varepsilon(N)$ depend on the norm, but the norm used is always clear from the context. It is generally impossible to compute $P_N^*(x)$ exactly, so we must consider estimates of $P_N^*(x)$. These estimates are produced by various computational algorithms and we have

*This work was partially supported by NSF grant GP32940X

Definition 1. *An algorithm A which produces an estimate $P_L(x)$ of $P_N^*(x)$ so that, as N and $L(N) \to \infty$,*

$$||f-P_L|| = \mathcal{O}(\varepsilon(N))$$

is called an optimal order L-parameter algorithm. The letter $M = M(A,N)$ denotes throughout the number of $f(x)$ evaluations required by A to compute $P_L(x)$. If $L = N$ and $M(A,N) = \mathcal{O}(N)$ then A is simply called an optimal algorithm.

The complexity of the algorithm is measured by M.

We denote the best approximation operator by T_N: $f(x) \to P_N^*(x)$ and we measure the complexity of T_N for a class C of functions by

$$M^*(N,C) = \inf_A \sup_{f \in C} M(A,N)$$

It is easy to believe (but not proved here) that M* cannot be less than $\mathcal{O}(N)$ for any interesting class of functions.

Our ideal objective is to show that M*=N for various norms (e.g., L_1, L_2 and L_∞), approximation forms (e.g., polynomials, splines) and classes of functions (e.g., $C^P[-1, 1]$, analytic in $|z| < 2$). Of course, we also wish to identify a corresponding optimal algorithm. We are able to do this in some cases and to come close in others. A significant conclusion derived from the results here is that asymptotically it is as easy to compute L_∞ approximations as L_2 approximations for most functions. A second significant conclusion is that, for a wide class of functions, piecewise polynomial approximations are no more complex to compute (even perhaps less complex) than ordinary polynomial approximations of comparable accuracy. We note that piecewise polynomials are much less complex to use than ordinary polynomials.

2. DISCRETIZATION

The first algorithm we consider is

Algorithm 1 (Discretization) Set $X = \{ih | i=0, 1, 2, \ldots, 1/h\}$, *evaluate* $f(x)$ *on* X *and then compute* $P_L(x)$ *as the best approximation to* $f(x)$ *on* X.

This algorithm is directly applicable to L_1, L_2 and L_∞ approximations. It was pointed out by Rice [73] that a minor variant algorithm is not very useful for smooth functions and that one obtains $M = N^p$ for the class $C^p[0,1]$. Since then, Dunham [74] has shown that if the end points 0 and 1 are included in X (as they are) then a better result holds. We have

Theorem 1. Consider the class $C^p[0,1]$, $p \geq 2$, *and polynomial approximation in the* L_1, L_2 *or* L_∞ *norms. Then discretization (Algorithm 1) is an optimal order N-parameter algorithm with*

$$M = N^{p/2}$$

Proof. Let $P_N(x)$ be approximation produced by Algorithm 1. Dunham [74] has shown that $||P_N - P_N^*|| = \mathcal{O}(h^2)$. We have then that $\varepsilon(N) = N^{-p}$ and $M = 1/h$ and we may eliminate h from the relation $h^2 = N^{-p}$ to obtain the conclusions stated.

Corollary. Algorithm 1 is an optimal algorithm for L_1, L_2 *or* L_∞ *approximation by polynomials for the class* $C^2[0,1]$.

3. LEAST SQUARES APPROXIMATION BY POLYNOMIALS

There appear to be two main algorithms for estimating least squares approximations by polynomials. For convenience we do least squares approximation with respect to the weight function $(1-x^2)^{-1/2}$ on $[-1,1]$. They are

Algorithm 2 (Gauss Quadrature for Fourier Coefficients). Estimate the coefficients

$$a_k^* = \int_{-1}^{1} \frac{f(x)\, T_k(x)\, dx}{\sqrt{1-x^2}}$$

by the Gauss quadrature formula:

$$a_k = \frac{2}{m} \sum_{i=1}^{m} f(\xi_i^{(m)})\, T_k(\xi_i^{(m)})$$

Thus a_k^* is the coefficient of the k-th Tchebycheff polynomial $T_k(x)$ in the Tchebycheff expansion of $f(x)$. The points $\xi_i^{(m)}$ are the m-point Gauss quadrature abscissa. The use of this algorithm and closely related ones is discussed in some detail by Rivlin [74, Section 3.5].

Algorithm 3 (Interpolation at the Tchebycheff points). Determine the polynomial $P_L(x)$ so that

$$P_L(\xi_i^{(L+1)}) = f(\xi_i^{(L+1)}) \qquad i = 1, 2, \ldots, L+1$$

It is well known that for P_L determined by Algorithm 3 we have

$$||f - P_L||_\infty \le ||f - P_L^*||_\infty\ (2/\pi \log L+1)$$

We note that if $m = L$, then the polynomials obtained by Algorithms 2 and 3 are the same (see Rivlin [74]).

Our first result on least squares is

Theorem 2 Consider the class $C^p[-1,1]$ $p \ge 3$ and least squares approximation by polynomials. Then Algorithm 2 is an optimal order N-parameter algorithm with

$$M = N^{\frac{p}{p-1}}$$

Proof: We restrict our attention to $m \geq N$ and we have from Rivlin [74, Theorem 3.12] that

$$||P_N^* - P_N|| < \sum_{j=1}^{\infty} \sum_{i=2jm-N}^{2jm+N} |a_i|$$

Now if $f(x) \varepsilon C^p[-1,1]$ we have that $|a_i| = \mathcal{O}(i^{-p})$ and we may estimate the inner sum, for some constant c, by

$$\sum_{i=2jm-N}^{2jm+N} |a_i| \leq \sum_{i=2jm-N}^{2jm+N} \frac{c}{i^p} \leq \sum_{i=(2j-1)m}^{(2j-1)m+2N} \frac{c}{i^p}$$

$$\leq \frac{c}{(p-1)[(2j-1)m]^{p-1}}$$

For $p \geq 3$ we then have that, for some constant c',

$$||P_N^* - P_N|| \leq \sum_{j=1}^{\infty} \frac{c}{(p-1)[(2j-1)m]^{p-1}} \leq \frac{c'}{m^{p-1}}$$

We now choose $m = M = N^{\frac{p}{p-1}}$ to obtain the correct order in the error $||f-P_N||$ and this concludes the proof.

We note that the previous result in Rice [73] corresponds to obtaining

$$M = N^{\frac{p}{p-1.5}}$$

and thus this sharpens that result. It seems likely that slightly more care in the proof would allow one to include the case p = 2, but then Algorithm 1 is already known to optimal for the class $C^2[-1,1]$.

Theorem 3. Consider the class $C^p[-1,1]$ and least squares approximation by polynomials. Then Algorithm 3 is an optimal order L-parameter algorithm with

$$L = M = N \sqrt[p]{\log N}$$

Proof. We have already noted that $||f-P_L||_\infty = \mathcal{O}(\varepsilon(L) \log L)$ and we also have that $||f-P_L||_2 \leq ||f-P_L||_\infty$. We have that $\varepsilon(L) = L^{-p}$ and $\varepsilon(N) = N^{-p}$. We claim that if $L = N \sqrt[p]{\log N}$ then $\varepsilon(L) \log L$ is $\mathcal{O}(N^{-p})$ because

$$(N \sqrt[p]{\log N})^{-p} \log (N \sqrt[p]{\log N}) = N^{-p}(\log N)^{-1} [\log N + 1/p \log N]$$

$$= N^{-p}(1+1/p)$$

This concludes the proof.

We see that Algorithm 3 uses fewer f(x) evaluations than Algorithm 2, but it does not result in an Nth degree polynomial.

The non-optimality in Theorems 2 and 3 arises from functions in $C^p[-1,1]$ where the Tchebycheff expansion coefficients a_k^* are the order of k^{-p}. These functions are rather special since we must also have

$\Sigma_{j=k}^{\infty} a_j^*$ the order of k^{-p}. Thus these functions have a very few large coefficients and the rest are comparatively negligible. The bulk of the functions in $C^p[-1,1]$ would seem to be covered by the next Theorem.

Theorem 4. Consider the subclass of $C^p[-1,1]$ which has $|a_k^*| = \mathcal{O}(k^{-p-1})$ and least squares approximation by polynomials. Then Algorithms 2 and 3 produce $P_N(x)$ with $||f-P_N|| = \mathcal{O}(N^{-p})$ and $M = N$.

Proof. We must, of course, take m = N in Algorithm 2 and L = N in Algorithm 3. We have already noted that the two algorithms produce the same polynomial in this case, so we may restrict our attention to Algorithm 2. If we repeat the proof of Theorem 2 with $|a_i| = \mathcal{O}(i^{-p-1})$ instead of $\mathcal{O}(i^{-p})$ we see that the final estimate turns out to be

$$||P_N^* - P_N|| \leq \frac{c'}{m^p}$$

and the choice m = M = N produces the specified order in the error which concludes the proof.

Note that Theorem 4 does not state that Algorithms 2 and 3 are optimal for the subclass considered. They are not optimal because one sees that

$$||f-P_N^*||^2 = \sum_{j=N+1}^{\infty} a_i^{*2} \leq c \sum_{j=N+1}^{\infty} \frac{1}{i^{2(p+1)}} \sim \frac{c}{N^{p+1/2}}$$

and hence $\varepsilon(N)$ is not N^{-p}.

There are classes of smoother functions where Algorithm 2 and 3 are optimal. We have

Theorem 5. Consider $f(x)$ analytic in a region containing $[-1,1]$ and least squares polynomial approximation. Then Algorithms 2 and 3 are optimal.

Proof. It is known that the hypothesis on f(x) implies that $|a_i| < c\rho^i$ for some constants $\rho < 1$ and c. Of course, we have again taken m = L = N in these algorithms. We use the same estimate as in the proof of Theorem 2 to obtain that

$$\sum_{i=(2j-1)N}^{(2j+1)N} |a_i| \leq \frac{c}{1-\rho} \rho^{(2j-1)N}$$

and hence we have

$$||P_N^* - P_N|| \leq \frac{c\rho}{(1-\rho)(1-\rho^2)} \rho^N$$

It is well known that $\varepsilon(N) = \mathcal{O}(\rho^N)$ and hence we have established that $||f - P_N|| = \mathcal{O}(\varepsilon(N))$ with M = N. This concludes the proof.

4. TCHEBYCHEFF APPROXIMATION BY POLYNOMIALS

We first note that Algorithm 3 (Interpolation at the Tchebycheff points) is equally applicable to Tchebycheff approximation and, in fact, we have

Theorem 6. Consider the class $C^p[-1,1]$ and Tchebycheff approximation by polynomials. Then Algorithm 3 is an optimal order L-parameter algorithm with

$$L = M = N \sqrt[p]{\log N}$$

The proof is essentially identical with that of Theorem 3.

It is well known that the best least squares approximation is asymptotically as good as the best Tchebycheff approximation for analytic functions. Thus we immediately obtain from Theorem 5

Theorem 7. Consider $f(x)$ analytic in a region containing $[-1,1]$ and Tchebycheff approximation by polynomials. Then Algorithm 3 is optimal.

We now turn to the most common algorithm for computing Tchebycheff approximations:

Algorithm 4. (Remes Algorithm). Take a large number (say 2N) of points in [-1,1] and apply Algorithm 1 to obtain the best Tchebycheff approximation on this discrete set. Then apply the Remes algorithm (see Rice [64]), with this as initial guess and use the Murnaghan and Wrench [59] procedure to locate local maxima. Once convergence is achieved within the specified tolerance, check the error curve for extraneous maxima that invalidate the approximation obtained. The check is performed by sampling the error curve at a number of points proportional to N.

This statement of the Remes algorithm is one useful in practice. It is known (Rice [69], Werner [62]) that the Remes algorithm is Newton's method for a particular set of equations. As such it has two weaknesses: It might converge to a local solution that is not a global one and we do not know the number of iterations required before quadratic convergence sets in. In fact, the latter number is unbounded on the set $f(x) \varepsilon C^p[-1,1]$. Its strength is that it is quadratically convergent. In Rice [73] we introduced some rather abstruse function classes in order to identify those f(x) where the Remes algorithm (Newton's method) behaves well. The fact of the matter is that one cannot identify such classes of functions with natural mathematical terms. The following definition allows us to make a more direct and intuitive presentation of the result.

Definition 3. Consider $f(x) \varepsilon C^3[-1,1]$ with $||f||_\infty \leq 1/2$. Let $P_N^{(i)}(x)$ be the approximation obtained by the algorithm at the ith iteration and set $\delta_i = ||P_N^*(x) - P_N^{(i)}(x)||_\infty$. We say that the Remes algorithm converges normally with constant α for $f(x)$ if

(i) $\delta_i \leq \alpha(\delta_0)^{2^i}$ (quadratic convergence)

(ii) the a posteriori check validates the approximation obtained (convergence to the global solution)

With this we may reformulate Theorem 4 of Rice [73] as follows:

Theorem 8. Consider the class of functions in $C^p[-1,1]$ $p \geq 3$ where Algorithm 4 converges normally with constant

$\alpha \leq \alpha_0$. Then, for Tchebycheff approximation by polynomials, Algorithm 4 is an optimal order N-parameter algorithm with

$$M = \mathcal{O}(N \log \log N)$$

This more direct reformulation allows us to give a simpler proof than the one previously outlined.

Proof. The initial calculation of $P_N^{(0)}(x)$ requires $\mathcal{O}(N)$ evaluations of $f(x)$. Each iteration of the standard Remes algorithm requires 4N evaluations (3N are for the Murnaghan-Wrench estimation of local maxima of the error curve). The number of iterations required is determined by the condition that $\delta_i \leq N^{-p}$. Since $\delta_0 \leq 1/2$ we find that $i = \log \log N + c$ is a sufficient number where c is constant depending on α_0 and p. The validation check requires a further $\mathcal{O}(N)$ evaluation of $f(x)$ and the total number required is $\mathcal{O}(N \log \log N)$ as claimed.

Note that while Theorem 8 asymptotically specifies fewer evaluations than Theorem 6 (or Theorems 2 and 3 for least squares), this relation does not hold for problems likely to occur in practice. With the optimistic assumptions that the initialization and checking only require N evaluations each and that 4 iterations are required (independent of N!) we find that the Remes algorithm leads to 18N $f(x)$ evaluations. The values of N where Theorems 2, 3 and 6 start to require more evaluations are, for $p = 4$, $N = 324$ (Theorem 2) and $N = 10^{45590}$ (Theorems 4 and 6).

In a similar manner we may establish

Theorem 9. Consider the class of functions analytic in a region containing [-1,1] where Algorithm 4 converges normally with constant $\alpha \leq \alpha_0$. Then, for Tchebycheff approximation by polynomials, Algorithm 4 is an optimal *order N-parameter algorithm with*

$$M = \mathcal{O}(N \log N)$$

Proof. The proof is the same as Theorem 8 except for bounding the number of iterations in the Remes algorithm. The requirement that $\delta_i \leq \rho^N$ (where $\rho < 1$ is associated with the size of the region of analyticity of f(x)) leads to i = log N + c (c = constant depending on ρ and α_0) as a sufficient number of iterations. The theorem now follows immediately.

5. PIECEWISE POLYNOMIAL APPROXIMATION

In our previous paper we proved that the spline projection operator of deBoor [68] is an optimal algorithm for $C^p[-1,1]$, for L_∞ approximation by piecewise pth degree polynomials with N knots. A recent adaptive approximation algorithm of Rice [75], [76] allows us to substantially enlarge the domain of functions where an optimal algorithm is known. We do not describe the algorithm here, but we do define a class of functions for which this algorithm is applicable.

Definition 4. The class $S_q^p[-1,1]$ of functions has the followning properities:

a) *Each $f(x)$ is bounded in the L_q norm on $[-1,1]$*

b) *Each $f(x)$ has a finite number of singularities*

$$s_i,\ i = 1, 2, \ldots, R$$

We set $w(x) = \prod_{i=1}^{R} (x-s_i)$

c) *$f^{(p)}(x)$ is continuous between the singularities*

d) *There are constants K and $\alpha > -1/q$ so that $|f^{(p)}(x)| \leq K|w(x)|^{\alpha-p}$ if $x \neq s_i$.*

e) *For any interval $[x,x+\rho]$ we let $F_p(x,\rho)$ denote the L_q norm of $f^{(p)}(x)$ on this interval. Let $E(x,\rho)$*

denote the error in the quadrature formula used by the adaptive approximation algorithm. This is typically a Gauss formula of precision p. There is a number $\lambda = \lambda(f)$ called the characteristic length so that if $\rho < \lambda$ we have

(i) $E(x,\rho) \leq K\, F_p(x,\rho)\rho^{p+1}$ *if* $F_p(x,\rho) < \infty$

(ii) otherwise $E(x,\rho) \leq K\rho^{1+\delta}$ *for some* $\delta > 0$.

There are three pertinent remarks to be made about this definition. The first is that S_q^p contains essentially <u>all</u> functions of practical interest in approximation. The second is that S_q^p is a subset of the functions involved in the work of Burchard [76]. Finally, the somewhat lengthly part (e) of the definition is included to ensure that the algorithm is computationally effective. We note that <u>none</u> of the previous algorithms have this feature and computationally effective versions of them must have <u>at least</u> one additional fact about f(x), a fact analoguous to the characteristic length. The typical example of such a fact is the actual numerical value of the norm of $f^{(p)}(x)$. These facts provide a priori bounds on the oscillations of f(x) and its derivatives.

The work of Rice [69a] and Burchard [76] shows that $\varepsilon(N) = N^{-p}$ for the class S_q^p and it has been shown by Rice [75], [76] that his adaptive algorithm achieves this degree of convergence. A simple inspection of that algorithm shows that the number of function evaluations is proportional to the number N of knots. The factor of proportionality is typically 8 or 10 although this would grow with larger p. These result imply

Theorem 10. Consider the class $S_q^p[-1,1]$ *and* L_q *approximation,* $1 \leq q \leq \infty$*, by piecewise polynomials of degree p. Then the adaptive approximation algorithm is optimal.*

REFERENCES

deBoor, Carl [68] deBoor, Carl, "On Uniform Approximation by Splines," J. Approx. Thy., 1 (1968), 219-235.

Burchard, H. G. [76] Burchard, H. G., "On the Degree of Convergence of Piecewise Polynomial Approximations on Optimal Meshes" II, to appear.

Dunham, C. B. [74] Dunham, C. B., "Efficiency of Chebyshev Approximation on Finite Subsets," J. Assoc. Comp. Mach., 21 (1974), 311-313.

Murnaghan, F. D. and J. W. Wrench [59] Murnaghan, F. D. and J. W. Wrench, "The Determination of the Chebyshev Approximating Polynomial of a Differentiable Function," Math. Comp., 13 (1959), 185-193.

Rice, J. R. [64] Rice, J. R., The Approximation of Functions, Vol. I, Addison Wesley, Reading, Mass. (1964), Chapter 6.

Rice, J. R. [69] Rice, J. R., The Approximation of Functions, Vol. II, Addison Wesley, Reading, Mass. (1969), Chapter 9.

Rice, J. R. [69a] Rice, J. R., "On the Degree of Convergence for Non-Linear Spline Approximations," in Approximations with Special Emphasis on Spline Functions (I. J. Schoenberg, ed.), Academic Press, New York (1969), 349-365.

Rice, J. R. [73] Rice, J. R., "On the Computational Complexity of Approximation Operators," in Approximation Theory (G. G. Lorentz, ed.), Academic Press, New York (1973), 449-456.

Rice, J. R. [75] Rice, J. R., "Adaptive Approximation," J. Approx. Thy., to appear.

Rice, J. R. [76] Rice, J. R., "An Algorithm for Adaptive Piecewise Polynomial Approximation," to appear.

Rivlin, T. J. [74] Rivlin, T. J., The Chebyshev Polynomials, John Wiley, New York (1974).

Werner, H. [62] Werner, H., "Die Contrucktive Ermittlung der Tschebyscheff-Approximierenden im Bereich der Rationalen Funktionen," Arch. Rat. Mech. Anal., 11 (1962), 368-384.

HENSEL MEETS NEWTON -- ALGEBRAIC CONSTRUCTIONS IN AN ANALYTIC SETTING

DAVID Y. Y. YUN

Mathematical Sciences Department
IBM Thomas J. Watson Research Center
Yorktown Heights, New York 10598

1. INTRODUCTION

Before the advent of Berlekamp's algorithm (Berlekamp [71]) for factoring polynomials over finite fields, the activity of factoring a polynomial over the integers was usually restricted to cases where the degree of the polynomial was quite small. Berlekamp's algorithm coupled with a lifting technique for bringing the factors back to the domain of interest (which is, in general, $Q[y_1, y_2, \ldots, y_v]$) provided a dramatic improvement in the range of polynomials that could be treated. This lifting technique originates from the Hensel Lemma of number theory and p-adic analysis (Hensel [13]).

> Hensel Lemma: Let p be a prime in Z and F(x) be a given polynomial in Z[x]. Let $G_1(x)$ and $H_1(x)$ be two relatively prime polynomials in $Z_p[x]$ such that $F(x) \equiv G_1(x)\,H_1(x) \pmod p$. Then for any integer $k > 1$, there exist polynomials $G_k(x)$ and $H_k(x)$ in $Z_q[x]$, where $q = p^k$, such that $F(x) \equiv G_k(x)\,H_k(x) \pmod q$ and $G_k \equiv G_1 \pmod p$, $H_k \equiv H_1 \pmod p$.

The constructivve proof of this lemma provides a method for lifting factors over a finite field step-by-step toward a larger subdomain of Z[x], while preserving the relevant properties, and eventually toward the desired result in Z[x]. In his 1969 paper, Zassenhaus [69] proposed an extension to Hensel's constructive method which achieved quadratic convergence.

<u>Zassenhaus</u> <u>Construction</u>: Let p, $F(x)$, $G_1(x)$, and $H_1(x)$ be given as before in the Hensel Lemma. For any integer $k > 1$, there exist polynomials $G_k(x)$ and $H_k(x)$ in $Z_q[x]$, where $q = p^{2^{k-1}}$, such that $F(x) \equiv G_k(x)\, H_k(x) \pmod q$ and $G_k \equiv G_1 \pmod p$, $H_k \equiv H_1 \pmod p$.

Since the modulus increases quadratically as k increases, the size of coefficients in G_k and H_k doubles each time toward a fixed bound of the integral coefficients of the desired factors of $F(x)$.

Musser [71] and Wang and Rothschild [73], utilizing these lifting techniques devised and implemented methods for factoring polynomials over the rational numbers, which were deterministic, constructive, and efficient. Other useful applications of these lifting processes in algebraic computations, such as the computation of polynomial GCD's, were discovered by Moses and Yun [73]. Later, Yun [74a,74b] presented and analyzed several algorithms based on these techniques.

Although these lifting processes are mainly algebraic in nature, we will show their close relationship to another powerful analytic method, namely the familiar Newton's method. Through this perspective we hope to simplify and enhance understanding of both approaches as well as to compare different techniques, discuss convergence, and derive new methods.

2. P-ADIC REPRESENTATION

We first introduce the notion of a "p-adic" representation of integers and polynomials. Given any prime p in Z, each integer a in Z can be considered to have a unique p-adic representation

$$a = \sum_{i=0}^{k} a_i p^i, \quad |a_i| \le p/2, \quad k \text{ is finite.}$$

Correspondingly, any polynomial $F(x)$ in $Z[x]$ has a unique p-adic representation

$$F(x) = \sum_{j=0}^{d} F_j x^j \qquad (F_j \text{ in } Z,\ F_d \neq 0,\ d \text{ is the hence finite})$$

$$= \sum_{j=0}^{d} \left(\sum_{i=0}^{k_i} F_{ij} p^i \right) x^j \qquad (\text{by putting each } F_j \text{ in p-adic form})$$

$$= \sum_{i=0}^{k} \left(\sum_{j=0}^{d} F_{ij} x^j \right) p^i \qquad (k = \max_i k_i,\ F_{ij} = 0 \text{ for } k_i < i \leq k)$$

where it is a series in p^i with coefficients which are polynomials in $Z_p[x]$.

Let $Z[[x]]$ be the space of finite p-adic series in the indeterminate x over the integers. It is easy to show that the domain of polynomials in x over the integers, $Z[x]$ is isomorphic to $Z[[x]]$. Thus, each finite p-adic series is in one-to-one correspondence to a polynomial in $Z[x]$ so that it can be used as a canonical representation of that polynomial. This view point leads naturally to the notion of "p-adic approximation" of polynomials in $Z[x]$. That is, any polynomial in $Z[x]$ can be approximated by a sequence of elements from $Z[[x]]$. The sequence is finite, i.e., the approximation will eventually be exact, since polynomials have bounded coefficients (k is finite).

As an example, let p=5. Then for

$$F(x) = 19x^5+5x^4-48x^2+31x-1$$

$$= (-1+(-1)\cdot 5+1\cdot 5^2)x^5+(0+1\cdot 5)x^4+0\cdot x^3$$

$$+ (2+0\cdot 5+(-2)\cdot 5^2)x^2+(1+1\cdot 5+1\cdot 5^2)x^1+(-1)x^0$$

$$= (-1x^5+2x^2+1x-1)5^0+(-1x^5+1x^4+1x)5^1+(1x^5-2x^2-1x)5^2,$$

the first term, $(-x^5+2x+x-1)$, is in $Z_5[x]$ which is contained in $Z[[x]]$ and is considered to be the "first order p-adic approximation" to $F(x)$; the first two terms, $(-x^5+2x^2+x-1)+(-x^5+x^4+x)\cdot 5$ which is in $Z_{25}[x]$ also contained in $Z[[x]]$, is considered the "second order p-adic approximation" to $F(x)$; all three terms which is in $Z_{125}[x]$ and, in this case, equal to the polynomial $F(x)$ is the "third order p-adic approximation" to $F(x)$. Both Hensel's and Zassenhaus' constructions are approximation methods in $Z[[x]]$ in essence where the desired results of the computation are given to more and more correct terms of their p-adic representations at each step of the constructions. We shall derive them using Newton's iteration formula over the appropriate domains in the following section.

3.0 NEWTON'S ITERATION

In its most readily recognizable form, Newton's method provides an iteration formula

$$x_{k+1} = x_k - f(x_k)/f'(x_k)$$

for finding a root of some given differentiable function $f(x)$. One familiar derivation of this formula is via the first order Taylor series expansion of f about some point x_k:

$$f(x) = f(x_k) + f'(x_k)\ (x-x_k)\ .$$

Assuming x is, or is sufficiently near, a root of f so that $f(x) = 0$, and solving for x, we have

$$x = x_k - f(x_k)/f'(x_k),$$

hence the Newton's iterative formula.

3.1 NEWTON'S ITERATION ON P-ADIC SPACE

Let f be a bivariate function on $Z[[x]]$, $f(G,H): Z[[x]] \times Z[[x]] \to Z[[x]]$. Then the first order bivariate Taylor series expansion of f about the point (G_k,H_k) is

$$f(G,H) = f(G_k,H_k) + (G-G_k)\ \frac{\partial f(G_k,H_k)}{\partial G} + (H-H_k)\ \frac{\partial f(G_k,H_k)}{\partial H}\ .$$

Again, if (G,H) is assumed to be a root or sufficiently near a root of f, then f(G,H) can be considered zero and we have

$$f(G_k,H_k) = -(G-G_k)\ \frac{\partial f(G_k,H_k)}{\partial G} - (H-H_k)\ \frac{\partial f(G_k,H_k)}{\partial H}\ .$$

Now consider a particular function f to be $f(G,H) = F - GH$ for some given F in $Z[[x]]$. (Note that the root of this function f, i.e., (G,H), will clearly be factors of the polynomial F.) Then, the above equation becomes

$$F-G_kH_k = f(G_k,H_k) = (G-G_k) \cdot H_k + (H-H_k) \cdot G_k .$$

Since there are two unknowns, G and H, in this case, we do not solve for them as in the derivation of Newton's formula, but we consider the differences as "updates",

$$\Delta G_k = G-G_k, \quad \Delta H_k = H-H_k ,$$

and we have the equation

$$F - G_kH_k = G_k \cdot \Delta H_k + H_k \cdot \Delta G_k \tag{1}$$

to solve at each iteration. The solutions for each equation are then used to update the current approximations to the desired zero of f by

$$G_{k+1} = G_k + \Delta G_k \quad \text{and} \quad H_{k+1} = H_k + \Delta H_k . \tag{2}$$

The convergence of Newton's method is guaranteed by a "good initial guess". Here we let (G_1,H_1) satisfying $f(G_1,H_1) = 0 \pmod p$ (i.e., $F \equiv G_1H_1 \pmod p$) be the given "initial guess" which is "good" in the sense that it approximates $F(G,H) = 0$. Note that G_1 and H_1 are in $Z_p[x]$ which is contained in $Z[[x]]$, so that the approximation procedure is carried out in the proper domain of f. Inductively, on the k<u>th</u> step, assume the point (G_k,H_k) has been found such that $f(G_k,H_k)$ approximates zero in the sense that

$$f(G_k,H_k) = F - G_kH_k \equiv 0 \pmod{q_k} \tag{3}$$

Here q_k denotes some power of p which will be specified later. The previous derivation indicates that the equation

$$G_k \cdot \Delta H_k + H_k \cdot \Delta G_k = F - G_kH_k$$

must be solved for the next approximation (G_{k+1},H_{k+1}) satisfying

$$f(G_{k+1},H_{k+1}) = F - G_{k+1}H_{k+1} \equiv 0 \pmod{q_{k+1}} \tag{4}$$

for some higher power of p, q_{k+1}.
We observe that it would certainly be wasteful to solve the equation for the updates, G_k and H_k, to its full accuracy in the entire domain $Z[x]$ while the next approximation G_{k+1}

and H_{k+1} are expected to be accurate only modulo q_{k+1} or in $Z_{q_{k+1}}[x]$. Thus it is sufficient as well as reasonable to solve the equation in $Z_{q_{k+1}}[x]$:

$$G_k \cdot \Delta H_k + H_k \cdot \Delta G_k = F - G_k H_k \pmod{q_{k+1}} \tag{5}$$

in a simpler domain $Z_{(q_{k+1}/q_k)}[x]$. This greater efficiency is achieved (due to (3) which implies q_k divides $F-G_kH_k$) by replacing (5) by

$$G_kA_k + H_kB_k = (F-G_kH_k)/q_k \quad (\text{mod } q_{k+1}/q_k) \tag{6}$$

to solve for A_k and B_k in $Q_{(q_{k+1}/q_k)}[x]$ and replace (2) by updating with current approximation $G_{k+1} = G_k + q_kB_k$ and $H_{k+1} = H_k + q_k A_k$. It can be proved that the solvability of this equation (6) depends only on the condition that G_1 and H_1 are relatively prime, which we do assume.

We will now specify q_k in two different ways to get the two desired algorithms. These choices of q_k, however, are by no means the only possible or interesting ones.

3.2 *ZASSENHAUS' CONSTRUCTION*

Let $q_k = p^{2^{k-1}}$ for k=1,2,... . Then by (6) the equation to solve at each iteration is

$$G_kA_k + H_kB_k = (F-G_kH_k)/p^{2^{k-1}} \quad (\text{mod } p^{2^{k-1}}) \tag{7}$$

with the updating formula being

$$G_{k+1} = G_k + p^{2^{k-1}} B_k \text{ and } H_{k+1} = H_k + p^{2^{k-1}} A_k \tag{8}$$

Expressions (7) and (8) give precisely the iteration formulas that accomplish the desired results of Zassenhaus' Construction stated at the beginning of this paper. A p-adic series approximation is achieved such that for any $n \geq 1$

$$G_{k+1} = G_1 + pB_1 + p^2B_2 + p^4B_3 + p^8B_4 + \ldots + p^{2^{k-1}} B_k$$

$$H_{k+1} = H_1 + pA_1 + p^2A_2 + p^4A_3 + p^8A_4 + \ldots + p^{2^{k-1}} A_k.$$

Since G_k, H_k, A_k and B_k are all in $Z_{q_k}[x]$ which is the space of p-adic series with $\log_p q_k = 2^{k-1}$ terms, each updating step (8) doubles the number of terms in the current p-adic approximation (G_k, H_k) of the desired root of $f(G,H)$. Hence the Zassenhaus' Construction exhibits quadratic convergence.

3.3 *HENSEL LEMMA*

On the other hand, let $q_k = p^k$ for k=1,2..., then, according to (6), the equation to be solved is

$$G_kA_k + H_kB_k = (F-G_kH_k)/p^k \pmod p.$$

But modulo p, $G_k \equiv G_1$ and $H_k \equiv H_1$ for all k. Thus, solving the equation

$$G_1A_k + H_1B_k = (F-G_kH_k)/p^k \pmod p \qquad (9)$$

for A_k and B_k in $Z_p[x]$ is the inner loop of the iterative procedure. The following is then used for updating

$$G_{k+1} = G_k + p^kB_k \text{ and } H_{k+1} = H_k + p^kA_k \qquad (10)$$

This derivation precisely establishes the validity and an inductive construction of Hensel's Lemma. This construction also constitutes a p-adic series approximation with

$$G_{k+1} = G_1 + pB_1 + p^2B_2 + p^3B_3 + \dots + p^kB_k$$

$$H_{k+1} = H_1 + pA_1 + p^2A_2 + p^3A_3 + \dots + p^kA_k \text{ for any } n \geq 1.$$

Since A_k and B_k are in $Z_p[x]$ which is the space of p-adic series with one term each, the updating step (10) clearly adds one more term to the current p-adic approximation (G_k, H_k) each time. Hence Hensel's Construction is only a linearly convergent approximation method.

4. ORDER OF CONVERGENCE VERSUS COMPUTATIONAL EFFICIENCY

Summarizing, we began by considering a Newtonian approximation method on the space of p-adic series isomorphic to Z[x]. For a particular function $f(G,H) = F-GH$ whose roots constitute factors of F, we derived both Zassenhaus' and Hensel's Constructions by choosing a certain sequence $\{q_k\}$ of powers of p. The two choices for $\{q_k\}$ result in two p-adic approximation methods that exhibit different convergence behavior, one quadratic and the other linear.

Even though these are not the only two interesting cases, an important question to pose is which method achieves better computational efficiency (e.g., in terms of the cost or number of single precision integer multiplications). For numerical computations the quadratically convergent Newtonian iterations are usually more efficient than linear methods (c.f. Traub [64]). As a result, the Zassenhaus' construction was used extensively for factoring integral polynomials even for multivariate polynomials (Musser [71] and Wang and Rothschild [73]).

Miola and Yun [74] made a detailed study and analysis of the algebraic algorithms of Hensel and Zassenhaus in 1974. Their finding, both from theoretical analysis and experimental data, showed that the cost for Zassenhaus' construction is always higher than that of Hensel's construction for achieving the same accuracy in the p-adic approximation of the results, except for the first step of the constructions (from modulo p to modulo p^2) which is identical for both methods. Although this result is somewhat surprising (and rather lengthy to derive), the basic reasons are not difficult to explain. From our Newtonian point of view, equations such as (1), (5), (6), (7), and (9) are the so called "inner loop" of each Newtonian iteration, and most of the computational effort has to be put here, since the "outer loop" of a Newtonian iteration such as (2), (8), and (10) usually involves only simple updating (e.g., adding). For most numerical Newtonian iteration procedures, the inner loops usually consist of function and derivative evaluations. But the solution of Diophantine polynomial equations is the main task to be performed in the inner loops of these algebraic constructions. Zassenhaus' construction requires solving (7) modulo $p^{2^{k-1}}$ for k=1,2,..., while Hensel's construction only needs to solve (9) in $Z_p[x]$ with G_1 and H_1 for each $k \geq 1$. Since $Z_p[x]$ is a smaller and more structured domain (Z_p is a field) and repeated use of G_1 and H_1 makes it possible to use some "preconditioning", these constitute the main reasons for Hensel's construction to be more efficient. (One preconditioning is to solve $G_1 a_i + H_1 b_i = x^i$ for $i=0,1,\ldots,d$ where $d \leq$ degree of F (= degree of G_1 + degree of H_1). Then any solution A_k and B_k for a particular right-hand-side of (9) can be expressed as scalar linear combinations of the a_i's and b_i's respectively). Thus we encounter a set of Newtonian iteration procedures where the quadratically convergent method turns out to be computationally more costly (i.e., less efficient) than the linearly convergent construction, because the cost of inner loop for each iteration is too high. This is another case among many others in symbolic and algebraic computations where simple-minded intuition from numerical algorithmic studies can lead us astray (Brown [71], Collins [66], Fateman [74], Gentleman and Johnson [73]). Thus the importance of studying the interrelationship between symbolic and numerical algorithms and understanding the connections between analytical and algebraic techniques can not be overstressed.

5. DISCUSSION AND CONCLUSION

In our derivation from the bivariate Taylor series expansion of f, we have chosen a particular function, f(G,H) = F-GH, in order to derive methods for lifting factors of polynomials. The p-adic approximation procedure derived from this function f is called "coupled p-adic construction", since the unknowns are coupled together through multiplication. Various applications of this type of construction in algebraic computations, such as polynomial greatest common divisor (Moses and Yun [73]), are possible and will be discussed elsewhere (Yun [74a,75]). The point to be made here is that this derivation of an algebraic algorithm from the analytic iteration point of view on appropriate algebraic domains is a very general technique. For example, if f is chosen to be f(G,H) = F-SG-TH for given F, S, T in Z[[x]], we can similarly derive the "decoupled p-adic construction" (so named because the unknowns are not coupled through multiplication). Again, there is a class of algebraic computations which can make use of this decoupled p-adic construction (Yun [74b,75]). Also if f is chosen to be a function of one variable:

$$f(G) : Z[[x]] \to Z[[x]] ,$$

then still another class of p-adic approximation methods can be derived. Among the applications are finding nth roots of polynomials, as well as power series reversions and reciprocal computations. Kung [74] was able to show that the method of computing reciprocals of power series derived from Newton's method achieves the computing time of $4n + \log_2 n$ where n is the number of terms required in the result. This is, currently, the best known time bound for reciprocal computations.

While we will not discuss details of applications here, we do wish to emphasize the generality of these algebraic construction methods, particularly when they are understood and unified with analytic concepts and techniques. Hopefully, this unified point of view sheds more light on the underlying structures, the generalities and restrictions, as well as the advantages and drawbacks of different (algebraic, analytic, or numerical) algorithms. Such a unification can be useful not only for developing symbolic systems and in devising and analyzing algebraic algorithms but also in the study of computational complexity as a whole.

REFERENCES

Berlekamp [71] Berlekamp, E. R., "Factoring Polynomials over Finite Field", *Bell System Technical Journal*, Vol. *46*, 1967, pp. 1853-1859.

Brown [71] Brown, W. S., "On Euclid's Algorithm and the Computation of Polynomial Greatest Common Divisors", *JACM*, Vol. *18*, No. 4, October 1971, pp. 478-504.

Collins [66] Collins, G. E., "Polynomial Remainder Sequences and Determinants", *Am. Math. Monthly*, *73*, Aug.-Sept. 1966, pp. 708-712.

Fateman [74] Fateman, R. J., "On the Computation of Powers of Sparse Polynomials", *Studies in Applied Mathematics*, Vol. *53*, No. 2, June 1974, pp. 145-155.

Gentleman and Johnson [73] Gentleman, W. M. and Johnson, S. C., "Analysis of Algorithms, A Case Study: Determinants of Polynomials", Proceedings of 5th Annual ACM Symposium on the Theory of Computing, Austin, Texas (1973). pp. 135-142.

Hensel [13] Hensel, K., Zahlentheorie, Goschen, Berlin and Leipzig, 1913.

Kung [74] Kung, H. T., "On Computing Reciprocals of Power Series", *Numer. Math.*, *22*, pp. 341-348, Springer-Verlag, 1974.

Miola and Yun [74] Miola, A., and Yun, D. Y. Y., "The Computational Aspects of Hensel-type Univariate Polynomial Greatest Common Divisor Algorithms", Proceedings of EUROSAM '74 (ACM SIGSAM Bulletin No. 31), Stockholm, Sweden, August 1974, pp. 46-54.

Moses and Yun [73] Moses, J., and Yun, D. Y. Y., "The EZ GCD Algorithms", Proceedings of ACM Annual Conference, Atlanta, August 1973, pp. 159-166.

Musser [71] Musser, D. R., "Algorithms for Polynomial Factorization", Ph.D. Thesis, Computer Sciences Department, University of Wisconsin, August 1971.

Wang and Rothschild [73] Wang, P. S., and Rothschild, L. P., "Factoring Multivariate Polynomials over the Integers", *ACM SIGSAM Bulletin* No. 28, December 1973.

Yun [74a] Yun, D. Y. Y., "The Hensel Lemma in Algebraic Manipulation", Ph.D. Thesis, Department of Mathematics, M.I.T., also project MAC Report TR-138, November 1973.

Yun [74b] Yun, D. Y. Y., "A p-adic Division with Remainder Algorithm", *ACM SIGSAM Bulletin*, Vol. *8*, No. 4, November 1974 (Issue No. 32), pp. 27-32.

Yun [75] Yun, D. Y. Y., "P-adic Constructions and its Applications in Algebraic Manipulation", in preparation.

Zassenhaus [69] Zassenhaus, H., "On Hensel Factorization I", *Journal of Number Theory*, Vol. *1*, 1969, pp. 291-311.

$O((n \log n)^{3/2})$ ALGORITHMS FOR COMPOSITION AND REVERSION OF POWER SERIES

Richard P. Brent
Computer Centre
Australian National University

H. T. Kung
Department of Computer Science
Carnegie-Mellon University

ABSTRACT

Let $P(s) = p_1 s + p_2 s^2 + \ldots$ and $Q(t) = q_0 + q_1 t + \ldots$ be formal power series.

The <u>composition</u> of Q and P is the power series $R(s) = r_0 + r_1 s + \ldots$ such that $R(s) = Q(P(s))$. The <u>composition problem</u> is to compute $r_0,\ldots,r_n$, given $p_1,\ldots,p_n$ and $q_0,\ldots,q_n$.

The <u>functional inverse</u> of P is the power series $V(t) = v_1 t + v_2 t^2 + \ldots$ such that $P(V(t)) = t$ or $V(P(s)) = s$. The <u>reversion problem</u> is to compute $v_1,\ldots,v_n$, given given $p_1,\ldots,p_n$.

The classical algorithms for both the composition and reversion problems require $O(n^3)$ operations (see, e.g., Knuth, Vol. 2). In this paper we describe algorithms which can solve both problems in $O((n \log n)^{3/2})$ operations. The techniques used to obtain our results are applicable to several other problems.

1. INTRODUCTION

Let k be a field, which contains an nth root of unity for every positive integer n. (For example, k could be the field of complex numbers.) Let p_i, q_i, $i = 0,1,\ldots$, be indeterminates over k, A the extension field $k(p_0,q_0,p_1,q_1,\ldots)$, and s, t indeterminates over A. Suppose that E and F are finite subsets of A and that we perform computations in the field A. Let L(E mod F) denote the number of operations necessary to compute E starting from $k \cup F$.

Let $P(s) = p_1 s + p_2 s^2 + p_3 s^3 + \ldots$ and $Q(t) = q_0 + q_1 t + q_2 t^2 + \ldots$ be formal power series over A. The <u>composition</u> of Q and P is the power series $R(s) = r_0 + r_1 s + r_2 s^2 + \ldots$ such that $R(s) = Q(P(s))$ is a formal identity. The <u>composition problem</u> is to compute $r_0,\ldots,r_n$, given $\{p_1,\ldots,p_n,q_0,\ldots,q_n\} \cup k$. Define

$$\mathrm{COMP}(n) = L(r_0,\ldots,r_n \bmod p_1,\ldots,p_n,q_0,\ldots,q_n)$$

Let $P(s) = p_1 s + p_2 s^2 + p_3 s^3 + \ldots$ be a formal power series over A. The <u>functional inverse</u> of P is the power series $V(t) = v_1 t + v_2 t^2 + v_3 t^3 + \ldots$ over A such that $P(V(t)) = t$ or $V(P(s)) = s$ is a formal identity. The <u>reversion problem</u> is to compute $v_1,\ldots,v_n$, given $\{p_1,\ldots,p_n\} \cup k$. Define

$$\mathrm{REV}(n) = L(v_1,\ldots,v_n \bmod p_1,\ldots,p_n).$$

The classical algorithms for both the composition and reversion problems require $O(n^3)$ operations (see, e.g., Knuth [71]), or $O(n^2 \log n)$ operations if the fast Fourier transform is used for polynomial multiplication as pointed out in Kung and Traub [74, Section 4]. In this paper we describe algorithms which can solve both problems in $O((n \log n)^{3/2})$ operations.

In another paper, Brent and Kung [75], we shall give a complete treatment of the subject, which will include the following:

(i) The proof that the composition and reversion problems are equivalent (up to constant factors) if $MULT(n) = O(REV(n))$, where $MULT(n)$ is the number of operations needed to multiply two nth degree polynomials.

(ii) Other algorithms requiring, e.g., $O(n^2)$ and $O(n^{1.9037})$ operations which do not use the fast Fourier transform and are faster for small n.

(iii) An algorithm which can evaluate the truncated functional inverse, i.e., $V_n(t) = v_1 t + v_2 t^2 + \ldots + v_n t^n$, at one point in $O(n \log n)$ operations, and its application to the root-finding problem.

2. PRELIMINARY LEMMAS

Let $P(s) = p_0 + p_1 s + \ldots$, $Q(s) = q_0 + q_1 s + \ldots$, $U(s) = u_0 + u_1 s + \ldots$, etc. be formal power series over A.

<u>Lemma 2.1</u>

<u>If $U(s) = P(s)Q(s)$, then</u>

$$\underline{L(u_0, \ldots, u_n \bmod p_0, \ldots, p_n, q_0, \ldots, q_n) = O(n \log n)}$$

Proof

Use the fast Fourier transform (see, e.g. Knuth [71, p. 441]). ■

Lemma 2.2

If $U(s) = P(s)/Q(s)$, then

$$L(u_0,\ldots,u_n \bmod p_0,\ldots,p_n,q_0,\ldots,q_n) = O(n \log n)$$

Proof

Use Lemma 2.1 and Newton's method as in Kung [74]. ■

Lemma 2.3

If $P(s) = p_1 s + p_2 s^2 + \ldots$, $R(s) = Q(P(s))$ and $D(s) = Q'(P(s))$, then

$$L(d_0,\ldots,d_n \bmod r_0,\ldots,r_n,p_1,\ldots,p_n) = O(n \log n).$$

(Here the prime denotes formal differentiation with respect to s.)

Proof

By chain rule, $R'(s) = Q'(P(s))\cdot P'(s)$. Hence $D(s) = R'(s)/P'(s)$, and the result follows from Lemma 2.2. ■

Lemma 2.4

If $P(s) = p_1 s + \ldots + p_m s^m$,

$Q(t) = q_0 + q_1 t + \ldots + q_j t^j$, where $m \le n$ and $j \le n$ and

$R(s) = Q(P(s)) = r_0 + r_1 s + \ldots$, then

$$\underline{L(r_0,\ldots,r_n \bmod p_1,\ldots,p_m, q_0,\ldots,q_j)}$$

$$\underline{= O(jm(\log n)^2)}.$$

Proof

We may assume that j is a power of 2. Write $R = Q_1(P) + P^{j/2} \cdot Q_2(P)$, where Q_1 and Q_2 are polynomials of degree $j/2$. During the computation we always truncate terms of degree higher than n.

The proof is by induction, so we can assume that $P^{j/4}$ is known. Thus, $P^{j/2}$ can be computed with $O(jm \log jm) = O(jm \log n)$ additional operations, and multiplication by $Q_2(P)$ also requires $O(jm \log n)$ operations. If $T(j)$ operations are required to compute R and $P^{j/2}$, then Q_1 and Q_2 may each be computed in $T(j/2)$ operations. Thus,

$$T(j) \le 2T(j/2) + O(jm \log n),$$

so

$$T(j) = O(jm(\log n)(\log j)) = O(jm(\log n)^2). \qquad \blacksquare$$

Lemma 2.4 can also be proved by using the fast evaluation and interpolation algorithms of Moenck and Borodin [72], but this method involves larger asymptotic constants and may have numerical stability problems.

3. THE COMPOSITION PROBLEM

Write $P(s) = P_h(s) + P_r(s)$, where $P_h(s) = p_1 s + p_2 s^2 + \ldots + p_m s^m$ and

$P_r(s) = p_{m+1} s^{m+1} + p_{m+2} s^{m+2} + \ldots$, for $m = \left\lceil \sqrt{\frac{n}{\log n}} \right\rceil$. Then

$$\begin{aligned} Q(P) &= Q(P_h + P_r) \\ &= Q(P_h) + Q'(P_h)P_r + \tfrac{1}{2}Q''(P_h)(P_r)^2 + \ldots . \end{aligned}$$

Let $\ell = \left\lceil \frac{n}{m} \right\rceil$. Since the degree of any term in $(P_r)^{\ell+i}$ is $\geq n+1$ for any $i > 0$,

$$Q(P(s)) = Q(P_h) + Q'(P_h)P_r + \ldots + \frac{1}{\ell!}Q^{(\ell)}(P_h)(P_r)^{\ell} + O(s^{n+1}).$$

This equality gives us the following algorithm for computing the first n coefficients of $R(s) = Q(P(s))$:

Step 1. Compute the first n coefficients of $W(s) = Q(P_h(s))$. By Lemma 2.4 with $j = n$ and m as above, this can be done in $O((n \log n)^{3/2})$ operations.

Step 2. Compute the first n coefficients of $Q'(P_h(s)), Q''(P_h(s)), \ldots, Q^{(\ell)}(P_h(s))$. By Lemma 2.3, it takes $O(n \log n)$ operations for each $Q^{(j)}(P_h(s))$. Hence the whole step can be done in $O(\ell\, n \log n) = O((n \log n)^{3/2})$ operations.

Step 3. Compute the first n coefficients of $P_r^2(s), P_r^3(s), \ldots, P_r^{\ell}(s)$.

Step 4. Compute the first n coefficients of $Q'(P_h(s))P_r(s), \ldots, \frac{1}{\ell!} Q^{(\ell)}(P_h(s))(P_r(s))^{\ell}$.

Step 5. Sum the results obtained from step 4.

It is clear that steps 3, 4 and 5 can be done in $O((n \log n)^{3/2})$ operations. Therefore, we have shown the following

Theorem 3.1

$\mathrm{COMP}(n) = O((n \log n)^{3/2})$.

4. THE REVERSION PROBLEM

Define function $f: A(t) \to A(t)$ by $f(x) = P(x) - t$. Suppose that $V(t)$ is the functional inverse of P. Then $P(V(t)) = t$. Hence $V(t)$ is the zero of f, and the reversion problem can be viewed as a zero-finding problem. We shall use Newton's method to find the zero of f; other iterations can also be used successfully. (See Kung [74] for a similar technique for computing the reciprocals of power series and also Brent [75, Section 13].) The iteration function given by Newton's method is

$$(4.1)\quad \varphi(x) = x - \frac{f(x)}{f'(x)} = x - \frac{P(x)-t}{P'(x)},$$

so we have

$$(4.2)\quad \varphi(x) - V(t) = x - V(t) - \frac{(P(V(t)) + P'(V(t))(x-V(t)) + \ldots) - t}{P'(V(t)) + P''(V(t))(x-V(t)) + \ldots} = \frac{P''(V(t))}{2P'(V(t))}(x - V(t))^2 + O(x - V(t))^3.$$

Suppose that the first n coefficients, $v_1, v_2, \ldots, v_n$, of $V(t)$ have already been computed. Let x be taken to be $V_n(t) = v_1 t + v_2 t^2 + \ldots + v_n t^n$. Then by (4.2)

$$\varphi(V_n(t)) = V(t) + O(t^{2n+2}).$$

Hence by computing the first $2n+1$ coefficients of $\varphi(V_n(t))$ we obtain the first 2n+1 coefficents of V(t). Hence by (4.1) and Lemmas 2.2, 2.3, we have

(4.3) $\mathrm{REV}(2n+1) \le \mathrm{REV}(n) + \mathrm{COMP}(2n+1) + O(n \log n)$.

Therefore, by (4.3) and Theorem 3.1 we have shown the following

Theorem 4.1

$\mathrm{REV}(n) = O((n \log n)^{3/2})$.

ACKNOWLEDGMENT

The authors want to thank J. F. Traub of Carnegie-Mellon University for his comments on the paper.

REFERENCES

Brent [75] Brent, R. P., "Multiple-Precision Zero-Finding Methods and the Complexity of Elementary Function Evaluation," these proceedings.

Brent and Kung [75] Brent, R. P. and H. T. Kung, to appear, 1975.

Knuth [71] Knuth, D. E., The Art of Computer Programming, Vol. 2, Addison-Wesley, Reading, Massachusetts, 1971.

Kung [74] Kung, H. T., "On Computing Reciprocals of Power Series," Numer. Math. 22, 1974, 341-348.

Kung and Traub [74] Kung, H. T. and J. F. Traub, "Computational Complexity of One-Point and Multipoint Iteration," in Complexity of Computation, edited by R. Karp, SIAM-AMS Proc., Vol. 7, American Mathematical Society, 1974, 149-160.

Moenck and Borodin [72] Moenck, R. and A. B. Borodin, "Fast Modular Transforms via Division, Conf. Record IEEE 13th Annual Symposium on Switching and Automata, 1972, 90-96.

ABSTRACTS OF CONTRIBUTED PAPERS

K-PARALLEL SEARCH TECHNIQUES

by

E. Arjomandi
D. G. Corneil
University of Toronto

This paper presents various K-parallel algorithms for searching an undirected graph. A very powerful technique used in efficient sequential graph theory algorithms is depth-first search. This pattern of search invokes an ordering on the edges of the graph. When utilizing K processors, it is very difficult to maintain this ordering if we allow more than one vertex or more than one edge from a vertex to be scanned at a given time. Thus efficient depth-first search seems to be inherently serial. In this paper we show that if our graph is sufficiently dense, breadth-first search techniques come very close to optimal. Techniques for searching sparse graphs are also presented.

ENTROPY MEASURES IN PROVING LOWER BOUNDS: A CASE STUDY

by

Ian Munro
University of Waterloo

One method of proving lower bounds on the time required to perform a particular task is to assign a measure to the state of the computation at any given time, and then bound the change that the measure can undergo in one time step. Often these entropy measures are so natural that we do not really think of them as such; there are, however, cases in which rather unlikely looking measures are just what is needed to show a bound. Although such examples can occur in Analytic Complexity Theory we illustrate our point by considering the problem of running multiple knockout tournaments in a minimal number of rounds. In particular, we show that $\log n + \log \log n + 2$ rounds are necessary and sufficient to run an n player double knockout tournament.

ON THE ADDITIVE OPTIMALITY OF FAST ALGORITHMS FOR MATRIX MULTIPLICATION

by

Robert L. Probert
University of Saskatchewan

A matrix multiplication algorithm which does not use the commutative law is represented by an ordered triple of flow graphs, $F = \langle G_1, G_2, G_3 \rangle$, called an addition flow representation. The additive cost of a flow representation of a particular algorithm to compute (m, n, p) products is the number of additions/subtractions used by the algorithm. Operations of rotation and reflection are defined on flow representations, and compositions of these operations on F are shown to yield representations of algorithms to compute (u, v, w) products where (u, v, w) is any symmetric permutation of (m, n, p). Using this technique, the additive complexities of symmetric problems are related by an additive symmetry theorem.

As an example application of additive symmetry, the seven-multiplication algorithm for (2, 2, 2) products communicated by Winograd is shown to be additively optimal over all fast algorithms (which do not use commutativity) for multiplying matrices of order two.

Finally, it is noted that this additive symmetry applies to any system of dual problems and not merely to matrix multiplication problems.

UPPER-BOUND TO THE TIME FOR PARALLEL EVALUATION OF ARITHMETIC EXPRESSIONS

by

David D. Muller and Franco P. Preparata
University of Illinois at Urbana-Champaign

Let E be an arithmetic expression involving n variables, each of which appears just once, and the possible operations of addition, multiplication, and division, requiring times τ_a, τ_m, and τ_d respectively. Then a constructively achievable upper-bound to the time required for parallel evaluation of E is $(\tau_a + \tau_m)\log n/\log \alpha + \tau_d$, where α is the positive root of the equation $z^2 = z + 1$. R. P. Brent (J.A.C.M., 21, 2, pp. 201-206) obtained the upper-bound $\lceil 4 \log_2(n-1) \rceil$ when $\tau_a = \tau_m = \tau_d = 1$, while the present result improves this to yield $2.88 \log_2 n + 1$.

PARALLEL EVALUATION OF DIVISION-FREE EXPRESSIONS

by

Franco P. Preparata and David E. Muller
University of Illinois at Urbana-Champaign

The problem of the parallel evaluation of division-free arithmetic expressions is investigated, under the assumption that a sufficiently large number of processors is available. A given arithmetic expression involving only addition, multiplication and $|E|$ distinct variables (a primitive expression) is constructively restructured so that the depth of the resulting computation tree is no greater than $\log |E|/\log \beta$, where β is the positive real root of the equation $z^2 = 2z + 1$, giving $1/\log_2\beta$ 2.0806... This shows that if the operations of addition and multiplication take unit time, E can be evaluated in at most $2.0806 \log_2|E|$ steps. We also consider a family $\{E_j\}$ of primitive expressions, where the computation tree T_j of E_j is recursively defined by $T_j = T_{j-3}(T_{j-3}T_{j-4} + T_0) + T_0$; E_j can be evaluated in j steps by our algorithm and $|E_j|$ grows as $c\beta^j$, for some constant c. We formulate the conjecture that the evaluation of E_j cannot be further sped-up by algebraic manipulations; this conjecture suggests that $2.0806 \log_2|E| - c'$ (c' a constant) is a lower-bound to the evaluation time of certain division-free expressions.

UPPER BOUNDS ON THE COMPUTATIONAL COMPLEXITY OF ORDINARY DIFFERENTIAL EQUATION INITIAL VALUE PROBLEMS

by

Arthur G. Werschulz
Carnegie-Mellon University

With few exceptions, past work in analytic complexity theory has centered on the problem of finding the zero of a nonlinear scalar function or operator. In this paper, we consider the problem of finding upper bounds on the number of function evaluations sufficient to solve a system of ordinary differential equations to within a given error criterion, ε, with one-step and multistep methods. Our main results are as follows:

(1) For any ε there is a unique choice of order and step size which minimizes the number of function evaluations.

(2) As ε decreases, this "optimal order" and the number of function evaluations both increase. As $\varepsilon \to 0$, both the optimal order and the number of function evaluations tend to infinity, but very slowly.

(3) As $\varepsilon \to 0$, the optimal order for multistep methods is less than the optimal order for one-step methods; moreover, numerical results indicate that the optimal multistep order becomes less than the optimal multistep order within a practical range of interest.

(4) As $\varepsilon \to 0$, the cost of the optimal multistep method is greater than the cost of the optimal one-step method; however, numerical results indicate that the optimal multistep method is cheaper for all ε within a practical range of interest.

A 6
B 7
C 8
D 9
E 0
F 1
G 2
H 3
I 4
J 5